BRIDGES OF WASHINGTON, D.C.

A HISTORY & GUIDE

BOB DOVER

THE History PRESS

Published by The History Press
An imprint of Arcadia Publishing
Charleston, SC
www.historypress.com

Front cover. top: Frederick Douglass Bridge. *Bryan Sayer*; *bottom*: Arlington Memorial Bridge. *Bryan Sayer.*

Back cover, top left: Potomac Aqueduct, with Key Bridge under construction in 1922. *Author's collection*; *top right*: William Howard Taft Bridge. *Bob Dover.*

All photos by Bob Dover between 2013 and 2024, unless otherwise noted.

First published 2026

Manufactured in the United States

ISBN 9781467170048

Library of Congress Control Number: 2025945972

Notice: The information in this book is true and complete to the best of our knowledge. It is offered without guarantee on the part of the author or The History Press. The author and The History Press disclaim all liability in connection with the use of this book.

CONTENTS

INTRODUCTION

Although bridges are best known for serving a major function in the transportation network of any city, they are much more than that. The record of their construction, destruction, and repurposing tells the story of the history of their city, including its original settlement, its growth, the rise and fall of its industries, and the evolution of transportation technologies. Because of their large size, they are a dominant part of our visual world, adding form and color to our landscapes and cityscapes. Also, by being accessible to pedestrians and bicycles, they support recreation while, at the same time, doing their part in reducing vehicle emissions.

In my books *Bridgespotting: A Guide to Bridges That Connect People, Places, and Times* and *Bridgespotting Part 2*, I discuss the historical trajectory in these other, mostly unnoticed uses of bridges throughout the United States and Europe. I document how, in the days before railroads and automobiles, bridges served a variety of community functions besides just moving people to the other side of the river. However, these other functions had largely ceased by the 1950s and 1960s as aesthetics, recreation, decoration, commemoration, and pedestrian uses took a back seat to speed. The *Bridgespotting* books also describe the common methods that cities have used to deal with their bridges that have deteriorated or become too small to accommodate the growth in traffic volumes. These bridges were usually in the way, occupying real estate that was needed for the replacement bridge, and were sitting on piers or towers that blocked use of the river by ships and boats. Also, to be preserved or repurposed, these bridges needed costly repairs. Therefore, these obsolete bridges were either removed or, even worse, simply left in place to rust.

But the books also document hundreds of examples, beginning in the 1980s and still accelerating today, of how communities and transportation planners have atoned for ugly, utilitarian, midcentury bridges by designing newer bridges as modern works of public art, complete with wide sidewalks connecting bike and hiking trail systems. There has also been enormous progress in the preservation of historic bridges, turning negatives into positives all over the United States by repurposing obsolete bridges to carry recreational trails and serve as the centerpiece of new parks.

Like most cities, the District of Columbia's past record of incorporating aesthetics, tourist attractions, recreational amenities, historic preservation, and other community-friendly features into its bridges has been a mixed bag. The District has not been a leader in this movement. Many other cities have been out in front.

As a result, I was initially skeptical of the idea of a travel guide to the bridges of Washington, D.C. Living most of my life in the District's suburbs, I had seen all of the major bridges before, many of them hundreds of times. When researching tourist and recreational bridges for the *Bridgespotting* books, I visited, photographed, and studied a small number of the District's bridges, focusing only on those that I felt were important enough to be included in a book that had a multinational scope. Because of this uncertainty, I began the project tentatively. I decided to make another quick pass, for due diligence, just to verify that current conditions are the same as I remembered them from my childhood, from decades of living and working in and near the city, and from my visits to research the *Bridgespotting* books. Very quickly, within the first two or three trips, I realized that Washington's bridges are not as I remembered them from twenty or more years ago. Instead, they are much better.

If you are visiting the District for a short sightseeing trip, you are probably going to concentrate on the museums and monuments of the National Mall. Maybe you will cross Arlington Memorial Bridge on a tour bus to visit Arlington National Cemetery, but that is probably the only bridge you are likely to see on a weekend trip. However, if you are visiting for a longer period, or maybe you are working in a temporary position in the District for a few months, then you will want to see and enjoy the city beyond the monuments.

Once you do that, the historic, aesthetic, and recreational attractions of the dozens of other bridges in the District will be unavoidable. If you plan to hike in its parks or ride on its bike trails, you will be traveling over and under bridges that represent more than two hundred years of history, ranging in importance from the mundane daily lives of the residents to nation-changing events. If you wish to explore lavish gardens or capture scenic photos of its waterways and monuments—bridges. If you wish to admire breathtaking public artworks or learn how the District has commemorated its native citizens who have influenced national culture—bridges. If you choose to walk through its historic neighborhoods to dine at out-of-the-way restaurants or enjoy the historic architecture—bridges.

Hopefully, these experiences will be enhanced by having learned something or been entertained by an amusing anecdote about each of the bridges. That is the purpose of this guide—to describe the physical and historical context to increase appreciation and enjoyment of the District's bridges by tourists, history buffs, hikers and cyclists, and others.

Although it was not at the vanguard of the modern bridge movement, the District has been catching up quickly in the past twenty years. From wider sidewalks and bike lanes being added as older bridges are rehabilitated to development of a regional bike trail system by interconnecting trails across multiple bridges, to the subtle lavender lighting of the stone-faced piers under the Francis Case Bridge, to the full-blown modernism of the superstructure arches on the new Frederick Douglass Bridge, the District has fully embraced twenty-first-century bridge design. The changes have turned the waterfronts into attractive destinations for tourism and recreation, something that had never before been the case. They have also improved the accessibility of the city on foot or by bicycle, which, of course, are the only ways to actually see and experience the city.

1
PHYSICAL SETTING OF THE BRIDGES

GEOLOGY AND TOPOGRAPHY

The types and sizes of bridges in any city are a function of a city's geomorphology. Literally translated as the "shape of the earth," geomorphology describes where there are hills, valleys, or bodies of water that may act as barriers to overland travel because the ground surface is too steep, rocky, muddy, or wet. These barriers will obviously include water bodies such as rivers, harbors, and marshes but may also include deep valleys with steep slopes. The shape and size of the barrier will dictate the size and type of bridge needed to cross it.

Washington, D.C., straddles the boundary between two distinct geologic regions, presenting two different types of geomorphology and, therefore, two different physical settings for its bridges. Like many major cities on the East Coast of the United States, the District sits directly on the Fall Line, which marks the boundary between the crystalline bedrock of the Appalachian Piedmont and the soft, unconsolidated sediments of the Coastal Plain. In the District, the Fall Line generally follows a southwest–northeast orientation, entering Northwest D.C. at Georgetown, passing south of the Smithsonian National Zoo, and exiting Northeast D.C. near Takoma Park, Maryland. Northwest D.C. sits on the Piedmont, while Southeast, Southwest, and Northeast D.C. sit on Coastal Plain sediments.

Rivers and streams on the Piedmont are relatively shallow and narrow freshwater streams that flow over bedrock rapids between steep-sided bluffs and cliffs. This type of environmental setting has several effects on bridges. The shallow water, narrow width, and rocky substrate mean that these rivers and streams were never navigable for shipping. Construction of low-profile bridges on piers and arches in these areas does not interfere with ships or barges, and the bridges do not need to have movable sections to allow passage of boats. However, the steep slopes present a barrier to overland travel. Vehicle traffic must wind its way down a steep, curvy slope to cross a ford or low-profile bridge over a narrow river or creek and then wind its way back to the top of the other side. This is inconvenient to the smooth flow of high-speed traffic. As a result, high-profile bridges are often constructed from bluff top to bluff top, rising more than one hundred feet above the stream and allowing traffic to cross the entire valley within a minute or two.

Another characteristic of freshwater rivers and streams on the Piedmont that affects the construction of bridges is the need to support infrastructure and industrial uses. As has happened at many cities on the Fall Line, the population growth and development of an urban center occurred on the Coastal Plain side of the line, where saltwater harbors supported shipping-based trade with other cities. However, those communities required fresh water for the residents as well as drops in elevation that could be used to power mills. The only sources for that fresh water and drops in elevation were on the Piedmont side of the Fall Line.

Because the District of Columbia straddles the Fall Line, it encompasses sections of freshwater rivers and creeks on the Piedmont, as well as sections of estuaries on the Coastal Plain. The Potomac River enters Washington from the northwest through the Potomac Gorge, a narrow fifteen-mile-long valley lined by bedrock cliffs known as the Potomac Palisades. In addition, a tributary to the Potomac, Rock Creek, originates on the Piedmont in Montgomery County, Maryland, and flows south over crystalline bedrock into the District. Therefore, Northwest D.C., situated entirely on the Piedmont, is where you will find relatively low-profile pier-and-arch bridges, the high-profile bluff top–to–bluff top bridges, and multiple canal and water supply bridges.

At Little Falls, the Potomac River reaches sea level and becomes an estuary on the Coastal Plain. The Anacostia River does not have a free-flowing freshwater section. The Northwest and Northeast Branches are nothing more than small creeks, rising on the Coastal Plain in the Maryland suburbs and converging about a mile northwest of Bladensburg to form the Anacostia. Both the Northwest and Northeast Branches reach sea level and become tidal upstream of this confluence, so the entire length of the

Anacostia from Bladensburg to the Potomac, a distance of about eight miles, is a tidal estuary situated on Coastal Plain sediments.

In the 1790s, the two ports of Georgetown and Alexandria on the Potomac were chosen for the establishment of the "Federal City," and the northern shore of the Anacostia was designated as the shipyard for the new U.S. Navy. Most of the city today is bounded by the Potomac on the west and the Anacostia on the south and east; therefore, these estuaries are the locations of the District's large-scale bridges. Because these estuaries are wide and shallow, the bridges were relatively long and constructed on piers. Also, because they blocked shipping access, the bridges were moveable, an issue that continued to complicate bridge design in the city well into the twenty-first century.

REENGINEERING OF WATERWAYS

In addition to the natural topography and geomorphology, the growth and development of the city was influenced by man-made improvements within and along the shores of both the Potomac and the Anacostia. Canals were constructed to allow transport of goods between the Potomac and Anacostia and between the city and the region to the west. In addition, the shipping channels and shorelines of both rivers were deepened to improve access for larger ships, and the dredged sediment was used as artificial fill material to convert swampy areas to developable land.

Three major reengineering projects created the current configurations of the rivers, waterfronts, and islands: the construction of the C&O Canal along the non-tidal part of the Potomac, the Tidal Basin on the tidal part of the Potomac, and the Burnham Barrier on the Anacostia. These projects completely reconfigured the entire length of the city's riverfronts, affecting the utility of existing bridges and spurring the need for additional bridges. The bridges associated with these areas all exist within the artificial landscapes created by these massive reengineering projects from the early nineteenth to the early twentieth centuries.

For a distance of 184 miles, the C&O Canal and its towpath hide among the trees at the base of the bluffs on the northern shore of the Potomac, within a few hundred feet of the river itself. Although the narrow canal has no effect on the larger bridges that cross the Potomac, small-scale bridges, many of which support access to recreation and historic tourist attractions

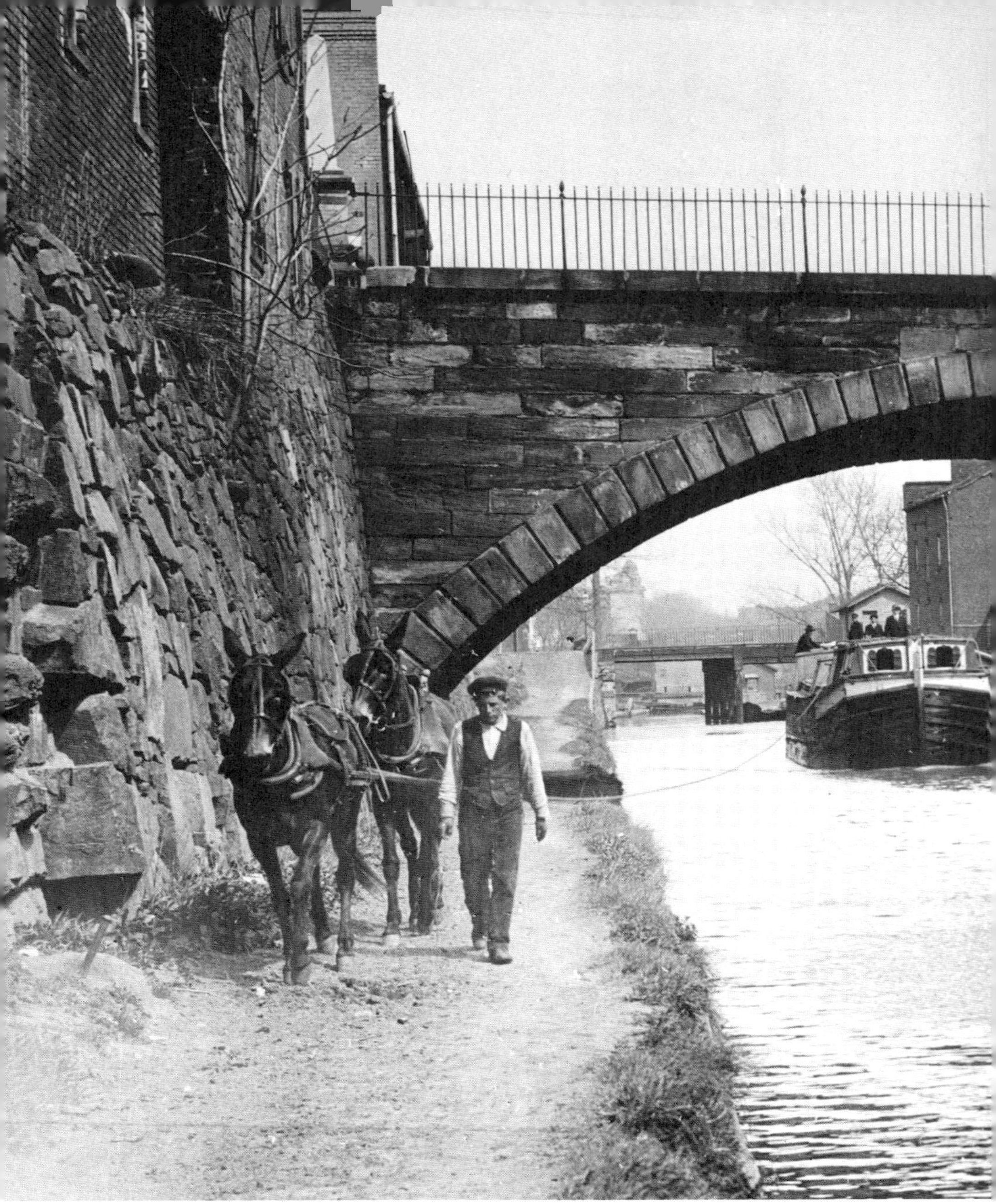

High Street Bridge during the heyday of the Chesapeake and Ohio Canal. *Library of Congress.*

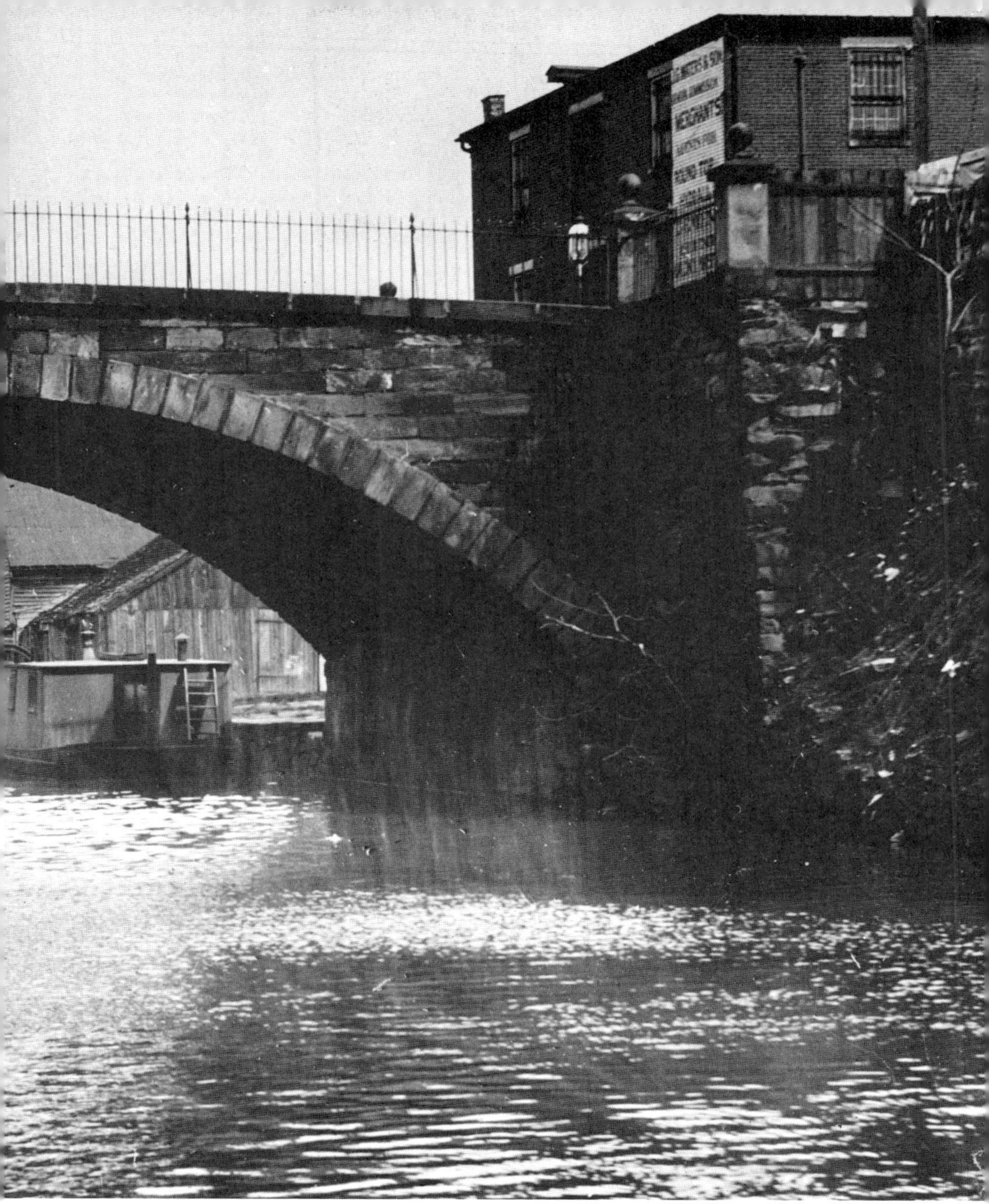

today, span the canal. The entire length of the canal is incorporated into the C&O Canal National Historical Park, operated by the National Park Service (NPS). As a result, much of the remaining canal infrastructure in the park, including locks, towpath, aqueducts to carry the canal over tributaries, lockhouses, foundations of former mills, and bridges have been preserved. These are all tourist and historical attractions, and the towpath is one of the most heavily used bike and hiking trails in the region.

A short distance below Little Falls and just above Georgetown, the Potomac River reaches sea level and becomes tidal. Shorelines of tidal estuaries are engineered in two ways that affect bridge construction. The first is that ships can access tidal estuaries, resulting in the development of port facilities. These include construction of docks, wharves, piers, and warehouses directly on the river, and housing and commercial buildings across the street. This kind of development usually occurs at what is known as the head of navigation of an estuary, because shippers want to be able to dock their ships as far inland as possible. Once this thriving commercial community is established, overland roads follow to facilitate trade, and these roads spur the need for bridges. In the District, this occurred at Georgetown, where early bridges were constructed northwest of the port to cross the river below Little Falls and southwest of the port to cross the river toward Alexandria. Today, these locations are Chain Bridge and Francis Scott Key Bridge. However, bridges between the ocean and the port must either be very high or have moveable spans to avoid interference with shipping. This is the case at Arlington Memorial Bridge, the Fourteenth Street Bridge complex, and the Woodrow Wilson Bridge.

The second effect of a tidal estuary is more problematic to urban development and transportation. Tidal estuaries are lined with tidal wetlands, which are great for the ecology but are also known as swamps. These swamps limit urban development because they do not provide a stable foundation for buildings and roadways, and they frequently flood. Because they are subject to floods, they are often associated with disease and pestilence. In addition to limiting urban development, clearing of surrounding areas for agricultural fields to feed the growing population causes erosion, which quickly leads to siltation in the harbor. Siltation reduces the depth of the harbor, interfering with shipping, which is why the community was constructed here in the first place.

To solve both of these issues, tidal wetlands in both the United States and Europe were engineered out of existence. In the tidal part of the Potomac, in the Anacostia, and in hundreds of other estuaries, shipping channels were

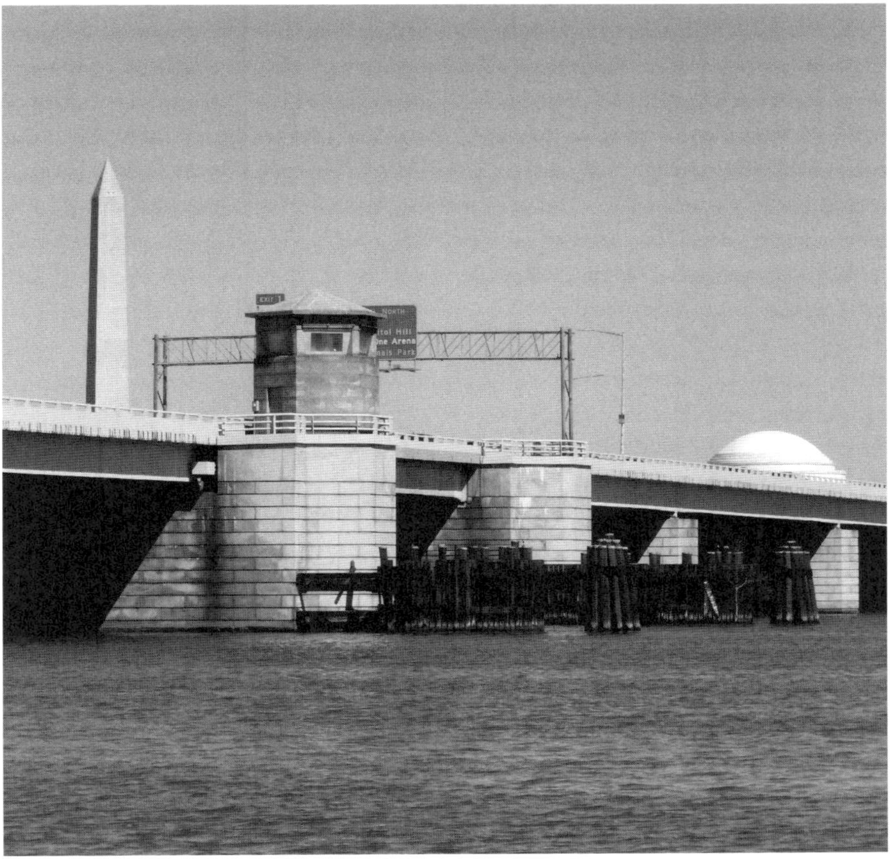

View of the operating house for the drawspan on the Fourteenth Street Bridge.

dredged to increase their depth, and the dredge material was used to raise the land elevation of the swamps to make developable land. Setting aside the fact that this was an ecological disaster, it was a win-win for the shipping industries and real estate developers.

On the tidal part of the Potomac, this was done to the entire eastern shore from Georgetown to the mouth of the Anacostia. In addition to maintaining the shipping channel to Georgetown, this created what are known today as the Potomac Embankment, the Tidal Basin, the Washington Channel and Southwest Waterfront, and Hains Point. Once those areas were turned into developable land, roadways and bridges soon followed, including Arlington Memorial Bridge, the Fourteenth Street Bridge complex, and the three bridges that enclose the Tidal Basin.

The Anacostia River has also been so extensively reengineered that it no longer resembles its original configuration. In addition to industrial development of the shorelines, the shipping channel was dredged to support the construction of ships at the Navy Yard and to maintain a channel to the port of Bladensburg. Similar to the dredging of the Potomac, the dredge material was used to build a complex of artificial islands known as the Burnham Barrier.

BRIDGES IN PARKS AND GARDENS

Most Americans, and even foreign tourists, know about what writer Wallace Stegner called "the best idea we ever had"—America's National Parks. What is not as well known is that the National Park Service manages about twenty different flavors of parks, of which only one flavor is designated with the name that is broadly known to the public as the "National Park." Other types of parks owned and operated by the NPS include National Battlefields, National Memorials, National Parkways, National Preserves, and a generic category of "Other Designations" that are simply named "Park," without any other adjective preceding it.

Although the District does not have an official National Park, much of its land area is occupied by parks designated as one or more of these other flavors of park. Specifically, the District is home to five National Historic Sites, twelve National Memorials, one National Monument, four National Parkways, and one National Scenic Trail. Many rankings of the accessibility and size of parks in U.S. cities, including U.S. News and World Report and the Trust for Public Land, list the District as having the largest land area of parks in the nation. Approximately twenty-four percent of the land area within the District's boundaries is designated parkland, most of it owned and maintained by the NPS. This includes Rock Creek Park, which, at 1,754 acres, is more than double the size of Central Park's 843 acres.

Most of these parks are small sites that are not associated with bridges. However, the George Washington Memorial Parkway, C&O Canal National Historical Park, and Theodore Roosevelt Island National Memorial, taken together, encompass both the western and eastern shores of the Potomac River, as well as the islands within the river and, therefore, are the setting for all of the bridges crossing the Potomac within the District. The George Washington Memorial Parkway includes the western shore of the Potomac, as well as Lady Bird Johnson Park at the western end of Arlington Memorial Bridge and Jones Point Park at the western end of the Woodrow Wilson Bridge. The C&O Canal National Historical Park occupies most of the eastern shore of the Potomac and is crossed by the larger-scale Potomac bridges, by street bridges in Georgetown, and by a series of small-scale footbridges. Roosevelt Island is situated in a key position between Georgetown on the north and the Mall on the south and is crossed by Theodore Roosevelt Bridge.

In addition to these parks and sites that have specific, formal designations, much of the land area of the District fits within the "Other Designation" category. These include the National Capital Parks-East (which includes Anacostia Park, River Terrace Park, and Kenilworth Park and Aquatic Gardens); Constitution Gardens, National Mall and Memorial Parks (which includes East and West Potomac Parks at the Tidal Basin); the White House; and Rock Creek Park, which includes the Rock Creek and Potomac Parkway, as well as a panoply of other noncontiguous small parks, former military sites, and gardens. Of these, Anacostia Park and the National Mall each cover a large land area that borders one of the rivers, and Rock Creek Park encompasses the entire length of Rock Creek within the District, as well as smaller parks along several tributaries. As a result, a substantial number of the District's bridges are located on or adjacent to parks managed by the NPS, even if they are not formally designated as "National Parks."

These parks owned and operated by the NPS comprise the Potomac riverfront and Rock Creek. They also encompass most, but not all, of the riverfront property on the Anacostia. The remaining Anacostia riverfront property is owned and operated as park or recreational space by other federal or city agencies. The National Arboretum on the Anacostia is owned by the U.S. Department of Agriculture. The city government operates The Fields at RFK Campus through Events DC (the city's convention and sports authority), and Kingman and Heritage Islands through the Department of Energy and Environment. A small segment of the Anacostia riverfront is part of the Washington Navy Yard, and Yards Park, which occupies the waterfront between the Navy Yard and Nationals Park, is privately owned but accessible for public recreation. Just outside the city boundary on the Anacostia, the Anacostia Trail and the Bladensburg Waterfront Park are operated by the Maryland–National Capital Park and Planning Commission. Each of these parks is associated with one or more of the Anacostia bridges, so they are an integral part of how tourists and locals can access both the bridges and parks.

Because most of the bridges are encompassed within the boundaries of federal and city parks, the bridges are, by definition, situated in close proximity to the tourist and recreational attractions in those parks. All of the best-known attractions of the District, including the museums and monuments of the Mall, the cherry blossoms at the Tidal Basin, Arlington National Cemetery, Nationals Park, and shopping and restaurants in Georgetown, are situated near parks that are within walking distance of one or more bridges. Even lesser-known but worthwhile attractions such as Kenilworth Park and Aquatic Gardens and Peirce Mill are situated in parks near bridges. Only one major outdoor attraction, the U.S. National Arboretum, is not directly associated with a bridge, but there are plans to rectify this with a new pedestrian bridge crossing the Anacostia.

Another result from the association of the bridges with the parks and gardens is aesthetics. Many of these bridges were constructed or otherwise influenced by landscape architects of the NPS. They are not only functional within the recreation network but also aesthetically attractive, enhancing both the natural scenery and the monumental architecture.

As early as the 1920s, and continuing in the twenty-first century, the NPS favored "rustic" designs for structures within its parks. Although concrete and steel are less costly and more durable construction materials than stone and wood, buildings and bridges constructed with stone and wood have a more natural appearance, giving park visitors the impression of being out in nature. Striking a balance between aesthetics and practicality led to park designs known as "parkitecture," which uses concrete and steel building materials for the structure but then hides these materials beneath stone facing to appear rustic. Another way to accomplish this look, which was done on the footbridges in Rock Creek Park, is to build with concrete but shape and texture the concrete into forms reminiscent of wood and stone structures.

Parkitecture can be seen throughout the District, not just within Rock Creek Park and its bridges. The George Washington Memorial and Baltimore–Washington Parkways are lined with stone-faced parapets instead of unsightly steel guardrails. Stone facing on concrete piers is used on many bridges, such as the Oregon Avenue Bridge over Pinehurst Branch. Even where concrete piers are left unfaced, such as on the Military Road Bridge, the abutments are decoratively faced with multicolored stonework.

One substantial difference between Rock Creek Park and the more famous Public Garden in Boston and Central Park in New York City is the style of landscape design. Both the Public Garden and the vast majority of Central

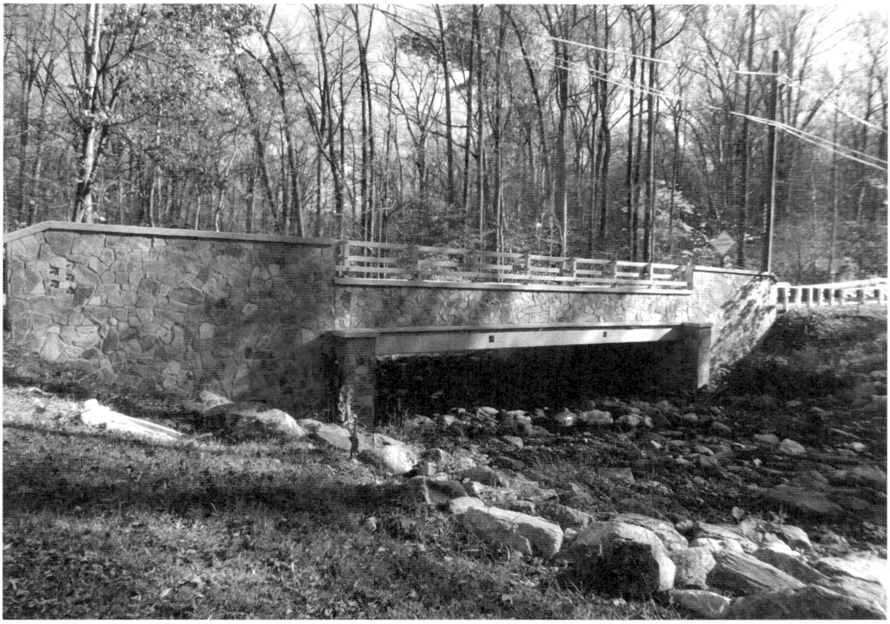

Although made of steel and concrete, the Oregon Avenue Bridge uses a stone façade as "parkitecture" in order to appear rustic.

Park are extensively landscaped with lawns, gardens, flower beds, cute artificial ponds and, crossing these cute ponds, tiny decorative footbridges. Central Park does have a large "rustic" area on its northern end that is not as manicured as the majority of the park, but the best-known areas consist of playing fields and gardens.

In contrast, the main part of Rock Creek Park, between the Maryland/District boundary and the zoo, is relatively wild and rustic. There are no manicured gardens in this area, only wooded hill slopes and narrow trails that require hiking boots. As a result, there are no cute bridges that serve a primarily decorative function. There are, however, several small concrete arch footbridges, constructed by the Public Works Administration (PWA) in the 1930s, carrying the hiking trails from one side of the creek to the other.

At the same time the National Park Service was influencing aesthetics on public spaces near the Mall and in Rock Creek Park, landscape architects were busy doing the same thing on privately owned estates. Following the lead of the landscape architects who had designed private gardens on estates in England, small-scale footbridges were constructed in private gardens both as decoration and to carry narrow trails through the scenery for recreation and enjoyment of nature. Dumbarton Oaks Park, not to be confused with the adjacent Dumbarton Oaks Gardens, was designed to offer a rustic stroll through nature and has a handful of unusual stone arch and clapper bridges connecting trails across the creek. A curved stone arch bridge serves as a driveway in the Tregaron Conservancy; a pair of arched wooden footbridges cross the Japanese Garden at the base of a waterfall in Hillwood Estates and Gardens; and a pair of concrete arches cross the fountain in the middle of Theodore Roosevelt Island.

Interestingly, as common as tiny, decorative, and functionless footbridges are in manicured gardens in other cities, they are relatively rare in the District, even in areas where they might be expected. Dumbarton Oaks Gardens is open to the public and offers fifty-three acres of manicured flower beds and winding trails surrounding the Dumbarton Mansion, yet it has no tiny decorative footbridges. Similarly, the National Arboretum covers 450 acres with displays of hollies, magnolias, azaleas, ferns, and conifers, including winding paths overlooking the Anacostia among exotic species displayed in its Asian Collections. Somehow, these amazing gardens were developed without including any cute decorative bridges. Kenilworth Park and Aquatic Gardens encompasses about 700 acres of ponds and tidal marshes, and while it has footbridges to carry hiking trails among the water lilies, none are decorative.

While each of these dozens of parks has its own internal trail system, they are also linked, through the sidewalks on the bridges, into regional-scale biking and hiking trail systems. As a result, the District probably has more large-scale bridges connecting to regional bike trails, thus supporting pleasant and safe biking for both recreation and commuting, than any other major U.S. city. For this reason alone, the city deserves a travel guide to its bridges.

2
HISTORICAL FRAMEWORK OF THE BRIDGES

As stated by Fred Emery in the Records of the Columbia Historical Society in 1938,

> *Washington has not the greatest bridges, nor the longest, nor the highest, nor the most costly, but what it lacks in massiveness of its waterway crossings is offset with historic environment along national lines.*

This is certainly still true of the District's bridges more than eighty-five years later. You will not find many panoramic postcard photos featuring the District's bridges as a prominent part of the cityscape. But all the same, these bridges have served as the stages for events of national importance that helped to shape our national identity. Armies have fought on them, and assassins have slipped across them in the dark of night. They are named for people who remind us of who we hope to emulate. Crowds have gathered on them to celebrate the ceremonial entries of our Founding Fathers and other leaders into the city and to mourn at the departure of our national heroes to their final resting place.

In addition to commemoration of national-scale leaders and events, these bridges also document the history of a growing city and how and where its people lived. As the population and traffic volumes grow, the need for new and larger bridges, using newly developed materials and technologies, also grows. Industries that require bridges come and go, and even the physical settings of the bridges change as riverfronts are reengineered and reimagined.

Bridges also degrade over time due to rotting wood, rusting of iron and steel, corrosion or brittleness of concrete, scouring at the base of piers, and uneven settlement of foundations in unconsolidated sediments. If caught in time during inspections, these structural issues are usually corrected through routine maintenance and periodic renovations or reconstructions. Not caught in time, these issues can lead to catastrophic failure, usually during a flood or ice floe event. When these situations happen, the original bridge may continue to live on through not-quite-forgotten ruins. Even if no remnants of the original bridge remain, the locations of these earliest bridges may still be tourist attractions for history buffs due to the bridges' roles in the development of the city or historic events that occurred there.

Combining the physical, economic, and political factors with the operation of 230 years of history, the bridges in the District can be categorized as follows:

- Early bridges connected the new Federal City to the Northern and Southern states and connected the two halves of the city on either side of Rock Creek.
- Infrastructure bridges allowed movement of bulk raw materials, manufactured products, agricultural products, and drinking water into and out of the city.
- Inner suburban bridges facilitated early real estate development west of Rock Creek and in Montgomery County, Maryland.
- Monumental bridges, constructed during the City Beautiful Movement as part of the McMillan Plan, returned large parts of the city to the original L'Enfant Plan of 1791 and transformed the city into an international landmark to rival Paris or London.
- Outer suburban bridges, constructed in the post–World War II period, accommodated the growing number of automobile-using commuters.
- Modern bridges enriched the community by combining aesthetics, commemoration, and recreation-friendly features.

This is a simplification, and there is a large amount of overlap between these categories, but this framework generally encapsulates the entire history of bridge construction in the District.

CONNECTING WASHINGTON, D.C., TO THE OUTSIDE WORLD

One unique feature of Washington, D.C., that makes it different from almost all other cities in North America and Europe, is that it is artificial. The city was not settled and developed based on its geographical advantages for industry, agriculture, or trade. Small ports existed at Alexandria, Georgetown, and Bladensburg in the eighteenth century, and these probably would have grown throughout the nineteenth and twentieth centuries. However, they never would have rivaled better harbors at the long-established cities of Baltimore, Philadelphia, New York, or Boston. The District may not have developed into the megalopolis that it is today if it had not been specially selected by George Washington in 1790 to be the site of the Federal City and then developed according to a grand design by Pierre L'Enfant.

Because it was the Federal City for an incipient nation, the District required immediate connection to the industrial, financial, and population centers of that nation. At its founding, the population distribution in the early United States was heavily weighted toward the Northeast. Baltimore, Philadelphia, New York, and Boston were all major ports, population centers, and hubs of finance and industry. All of these cities are situated northeast of Washington, implying that the most heavily used traffic route into and out of the city would have been on the northeastern side. This route was known as the Baltimore Pike, and a bridge was constructed across the Eastern Branch of the Potomac (now known as the Anacostia) at the port of Bladensburg, which is in Maryland about one mile north of the District border, as early as 1770. L'Enfant recognized the importance of this connection in his 1791 Plan by orienting three grand avenues—eventually named New York, Rhode Island, and Maryland Avenues—radiating southwest from this bridge to the President's House, the Capitol, and Georgetown.

L'Enfant also planned for bridges crossing the Eastern Branch to connect to areas south of the city. Shortly after the founding of the city, the Washington Navy Yard was developed on the Eastern Branch directly south of the Capitol. To house Navy Yard workers in a suburb on the opposite side of the Eastern Branch, bridges were constructed at the foot of Pennsylvania Avenue in 1804 and at Eleventh Street adjacent to the Navy Yard in 1820, becoming the third and fifth major bridges connecting the District to its surroundings.

The first bridge to cross the Potomac was at the current location of Chain Bridge, northwest of Georgetown. Constructed in 1797, this bridge crossed the river at the location of a flourishing industrial settlement at Pimmit Run, but it also allowed business interests in Georgetown to reach markets northwest of the new city without needing to pass through Alexandria. Similarly, the Long Bridge, constructed in 1809, crossed the Potomac on the southwestern corner of the city, creating the shortest route between the governmental center and the port of Alexandria without going through Georgetown.

In addition to the waterways bordering the new Federal City, the chosen location for the city was split in two by the incised valley of Rock Creek. Georgetown is situated west of Rock Creek. However, the city planners decided to construct the government buildings, including the Capitol and the President's House, east of Rock Creek. Even before the District was selected in 1790 as the new Federal City, a bridge was constructed across Rock Creek in 1788 at the location now occupied by M Street, followed quickly by one at the current location of K Street. After the selection, the 1791 L'Enfant Plan proposed two additional bridges connecting Georgetown to the government center.

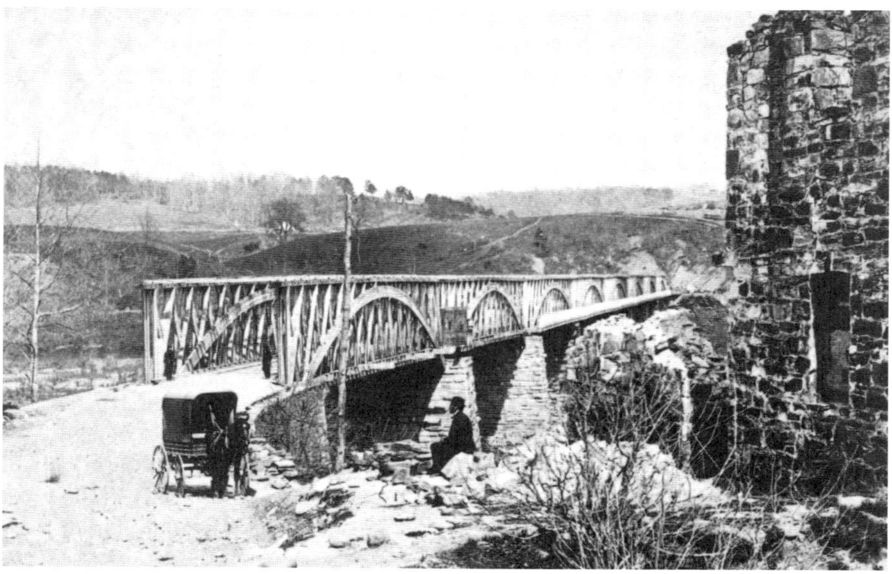

View of one of the nineteenth-century bridges at Little Falls, now known as Chain Bridge. *Library of Congress.*

Even though none of these early bridges remain, their influence on today's city is still found in the roadway patterns and the locations of parks and neighborhoods. Each bridge was involved, to some extent, in the events of August 24, 1814, when the British entered Washington, D.C., and burned the President's House and Capitol. As a result, their locations may be visited by tourists interested in history today. All six locations have newer bridges at or near the original crossings, including the Bladensburg Road Bridge, the John Philip Sousa Bridge at Pennsylvania Avenue, the Eleventh Street Bridge at the Navy Yard, and more recent iterations of the Long Bridge, Chain Bridge, and M Street Bridge.

INFRASTRUCTURE BRIDGES

Shortly after bridges on the outer perimeter of the city were constructed to provide connections to the outside world, D.C.'s population started to grow, and industry developed. Both required additional bridges to move fresh water, coal, building stone, and food into the city.

The earliest non-road infrastructure constructed to bring materials and products into the city was the C&O Canal. Not a bridge in itself, the canal did require bridges, in the form of aqueducts, to be constructed wherever its route crossed a tributary entering the Potomac River from the east. In places where there were mills or other facilities situated between the canal and the river, small bridges crossed the canal to connect these to the city. Today, the oldest original bridge in the District, constructed in 1831, crosses the C&O Canal in the heart of Georgetown.

Two early large-scale bridges were constructed to bring drinking water to the growing population. Within the city, the Potomac and Anacostia Rivers are both salt water, and Rock Creek and the Northwest and Northeast Branches of the Anacostia are each too small to be substantial fresh water sources. However, the non-tidal Potomac just a few miles upstream of the District is a substantial freshwater river and an ideal water source. In the 1850s, the Washington Aqueduct was constructed to connect this water source to the center of the city. This necessitated constructing bridges to carry this water across Cabin John Creek a few miles northwest of the District boundary and across Rock Creek at what is now Pennsylvania Avenue. Although their primary purpose was water transport, both bridges were also developed to carry vehicles and are busy traffic bridges today.

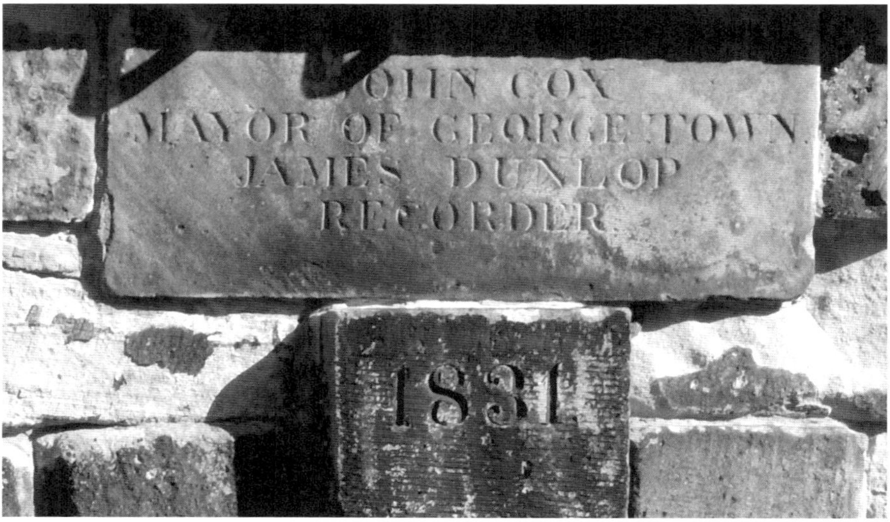

The names of the mayor and recorder of Georgetown and 1831 construction date on the Wisconsin Avenue Bridge at the C&O Canal.

Later in the nineteenth century, railroad bridges across both the Potomac and Anacostia were constructed. Because railroad bridges are constructed by privately owned companies for their own use, they are usually not special or attractive, which is the case with the District's three remaining railroad bridges. However, they are historic and appeal to bridge fanatics who are drawn to study and document early engineering technologies.

INNER SUBURB BRIDGES

As the population grew and improved transport methods were invented, the well-to-do of the inner city began to move out to the suburbs. In the District's early days, when the city consisted of a small core surrounding the Capitol and President's House, the suburbs meant the area on the other side of Rock Creek, in Northwest D.C. and Montgomery County, Maryland. This phase of city expansion, which continued from the early 1800s into the 1930s, led to multiple road bridges constructed across the Rock Creek Valley and its tributaries.

Some of the roadways crossing Rock Creek were narrow roads that wound down through the trees and rocks on one slope of the valley,

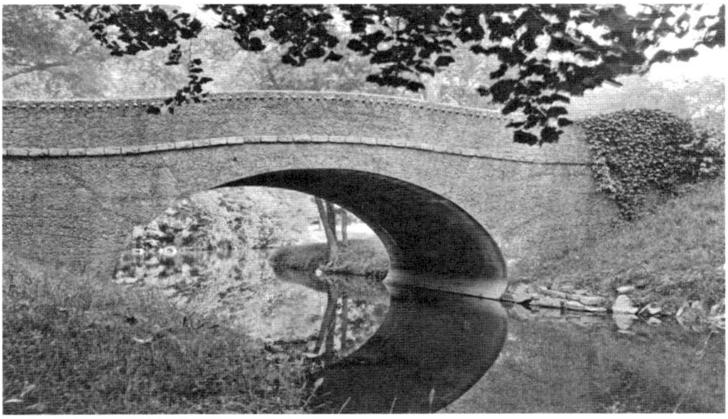

The decorative Pebble Dash Bridge carried Beach Drive across Broad Branch near Rock Creek from 1902 to 1957. *From the collection of John DeFerrari.*

crossed Rock Creek on a short, low-profile bridge, and then climbed back to the top on the other side. These roads, only a few short blocks from the busiest urban avenues, are the main access points for tourists, hikers, bicyclists, and other visitors to Rock Creek Park today. However, while they are scenic and used for recreation and tourist access, they were not suitable for twentieth-century commuters, most of whom cared more about traffic flow than scenery. Also, because they were hilly and winding, they interrupted the distinct radial spoke-and-wheel pattern of L'Enfant's grand design for the city's layout.

To maintain the straight lines of the L'Enfant Plan and provide faster, more direct routes across the park, a series of longer and higher bridges stretching from bluff top to bluff top was developed in the early twentieth century. This required constructing the world's highest and widest arch bridges at the time, breaking many records held by European bridges. A drive on the Rock Creek and Potomac Parkway, along the axis of the southern half of Rock Creek, passes beneath a series of these enormously impressive and historically important bridges.

In the case of these bridges constructed by land speculators to attract homebuyers to the other side of a creek or river, the developers had a vested interest in making those bridges both pedestrian-friendly and visually appealing. This opens up the real estate market to purchasers who may not have a car and entices customers by providing a pleasant journey to and from work every day. The Taft, Klingle Valley, Dumbarton, Ellington, and Piney Branch Bridges crossing Rock Creek and its tributaries were all constructed to lure homebuyers to what were then the suburbs of Northwest D.C. All of these bridges have wide sidewalks and are elaborately decorated with statues and carved monuments.

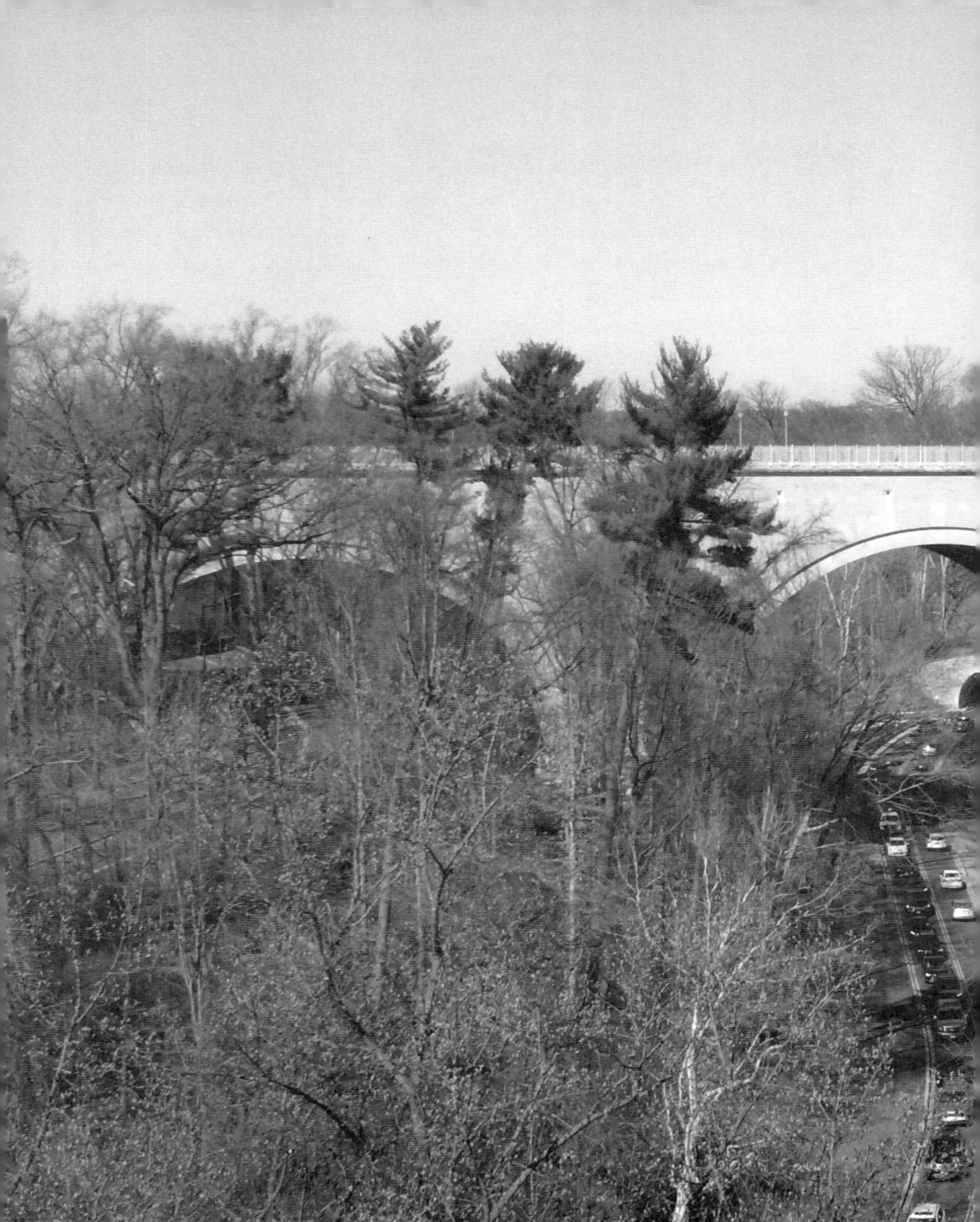

View of Duke Ellington Bridge and the Zoo Tunnel taken from the sidewalk on the William Howard Taft Bridge.

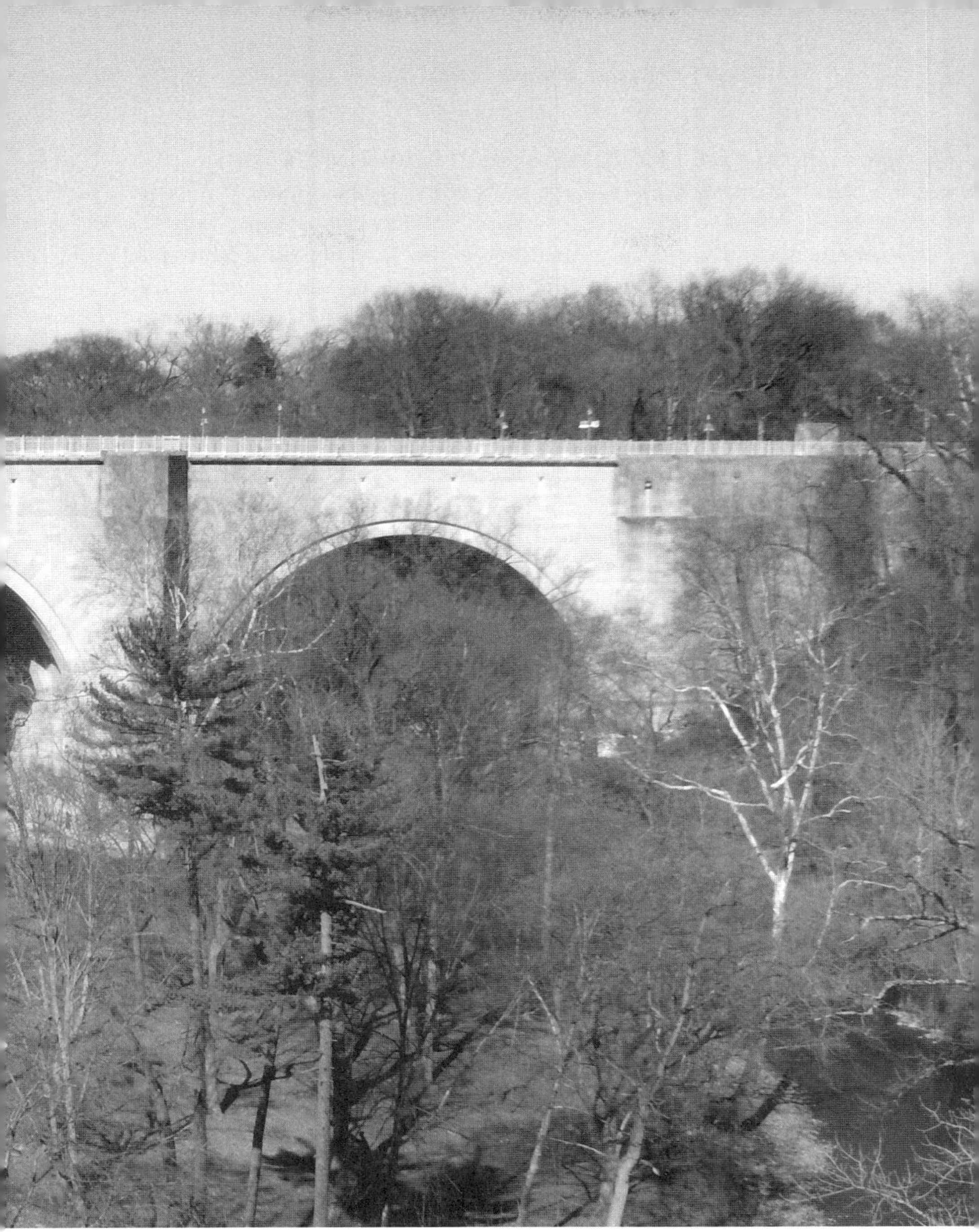

CITY BEAUTIFUL MOVEMENT AND McMILLAN COMMISSION

Entire books have been written about the City Beautiful Movement of the early twentieth century, and the District was a prominent focus. The City Beautiful Movement grew out of the development of landscape architecture, as exemplified by Frederick Law Olmsted and others in the second half of the nineteenth century. Originally practiced on privately owned gardens of the well-to-do in England, the movement influenced landscape architecture and led to the beautification of public areas in Boston's Public Garden in 1837 and in New York City's Central Park in the late 1850s. The idea was further developed to incorporate integration of aesthetic form and function in the years following the success of the World's Columbian Exposition in Chicago in 1893.

The height of the City Beautiful Movement, from 1900 to 1910, coincided with the establishment of the Senate Park Commission chaired by Senator James McMillan, head of the Senate District of Columbia Committee. McMillan's senior aide, Charles Moore, was connected to Glenn Brown, executive secretary of the American Institute of Architects (AIA), as well as other members of the AIA. The AIA, headquartered in Washington, D.C., was effectively a lobbying group for architects and, therefore, had an intense interest in development of public buildings and spaces in the city. Brown chose to hold the 1900 AIA annual convention in the District, to coincide with the city's one hundredth anniversary. The theme of the convention was "The Grouping of Government Buildings, Landscape, and Statuary."

At this convention, the architects criticized the existing plans of the Army Corps of Engineers and Office of Buildings and Grounds and argued for a return to the L'Enfant Plan. Moore used the convention to establish a connection between the AIA and McMillan, and McMillan agreed to work with the Senate to establish the Senate Park Commission in 1901. The three commissioners included Daniel Hudson Burnham, Charles Follen McKim, and Frederick Law Olmsted Jr. Burnham had designed the 1893 Columbian Exposition in Chicago and was an early advocate of the City Beautiful Movement. McKim was head of the McKim, Mead, and White architecture firm, and Olmsted was, along with his father, a leader in landscape architecture. The sculptor Augustus Saint-Gaudens joined as a fourth commissioner a few months later.

The commissioners first gathered on the axis connecting the Washington Monument to the Capitol on April 15, 1901. They then embarked on a tour of London, Paris, Rome, Venice, and Vienna for seven weeks, with the objective of studying the architecture and layout of those cities so that the District could be developed into a similar world-class capital. When they returned, they developed models and plans and presented them to the public at the Corcoran Gallery on January 15, 1902.

The McMillan Commission was responsible for the development of most of the major tourist areas visited in the District today. The entire National Mall, lined with the Smithsonian museums, the National Gallery of Art, the Lincoln Memorial, the parks and monuments surrounding the Tidal Basin, and most of the other memorials, are all the direct or indirect result of the McMillan Commission.

The McMillan Commission and City Beautiful Movement had two effects on the bridges that attract tourists in the twenty-first century. The first was the development of parkland on top of dredged fill material south of the Washington Monument, at the western end of the Mall and surrounding Twining Lake, now known as the Tidal Basin. Once these former swamp areas had been made into buildable land, the McMillan Commission proposed developing the new land into the series of parks that now line the eastern bank of the Potomac. Although some of the associated bridges, such as Arlington Memorial in 1923 and Kutz Bridge in 1943, were not constructed until later, they were a direct result of new land and park areas created by the McMillan Commission.

The second effect was the form and decoration of new bridges throughout the city. The years prior to World War II were a period of rampant decoration. Users of the bridges traveled on foot, streetcar, or other slow-moving vehicles, so they had time to view and appreciate sculptures, plaques, and other informative and commemorative ornamentation. The oldest bridge in the city, the High Street Bridge crossing the C&O Canal, displays the names of the mayor and recorder of the city of Georgetown on a stone plaque, as well as the construction date of 1831 engraved into the keystone of the arch. Also, many of the District's bridges are named and decorated to memorialize prominent former citizens, including presidents, composers, and civil rights icons. Even if these commemorations were added years or decades after a bridge was constructed, there is usually a plaque somewhere documenting the commemoration. These can be found on the Taft, Ellington, Glover, Theodore Roosevelt, and Key Bridges, among others.

Although the McMillan Commission was not directly responsible for the construction of the Dumbarton, Taft, Piney Branch, and other bridges outside of the monumental core of the city, the City Beautiful Movement did influence the form and decoration of these and other bridges throughout the city. Life-sized sculptures of lions and eagles can be found on Taft Bridge, bison on Dumbarton Bridge, and tigers on the Piney Branch Bridge. Enormous allegorical statues welcome drivers onto Arlington Memorial Bridge, and more subtle allegorical reliefs carved into stone monuments on the Duke Ellington Bridge celebrate the modes of transportation that were "modern" in 1935. Ornate medallions depicting eagles are found on the closed spandrel of Arlington Memorial Bridge, and intricately carved monograms of the initials "DC" are worth seeing on the spandrel of the Dumbarton Bridge.

COMMUTING BRIDGES

Following World War II, and especially as part of the development of the interstate highway system from the 1950s through the 1970s, bridge design in the United States became completely severed from the connection it once had with its communities. Bridges had one purpose: to move as much traffic as possible as quickly as possible from the suburbs into the city center in the morning and to reverse this pattern in the late afternoon. There was no need for pedestrian sidewalks, no thought of decoration or commemoration, and no consideration of the quality of life in the neighborhoods that were bulldozed to make way for highways or for residents who had to live in the shadow of an elevated highway.

On a national scale, this happened because, prior to passage of the National Environmental Policy Act of 1969, the public had no voice in federally funded construction and development projects such as bridges. The state and federal agencies constructing the bridges were not required to engage local residents in decisions related to bridge location, size, or design. The massive construction of the interstate highway system across the entire nation from the 1950s through the 1970s was completed by the federal government prior to the legal requirement to involve local communities, and unfortunately, this shows. In community after community, historic neighborhoods and viewscapes were destroyed by the construction of elevated highways with large-scale unattractive plate-girder and concrete-girder bridges with no consideration for aesthetics or recreational features.

On a local scale, bridge construction in the District was even more disconnected from the community than most other cities. From 1800 to 1973, there was no locally elected city government that had the responsibility and authority to develop and operate a road network and bridges. There were periods of direct congressional control, rule by presidentially appointed commissioners, and a brief, limited experiment with the election of a mayor in the 1870s. The current city government structure was developed through the establishment of the Home Rule Act in 1973, and it took even longer for residents in the lower-income neighborhoods of the city to be given a voice in city affairs. Even today, legislation passed by the city council is still subject to congressional approval, and any construction in the city is subject to the Building Heights Act of 1910 and approval by the U.S. Commission of Fine Arts.

As a result, the vast majority of the District, including the government buildings, monuments, parks, road network, and the associated bridges was developed either by representatives elected by constituents in faraway states or by politically appointed commissioners. None of these individuals was in any way accountable to the residents of the city, so it is no surprise that residents were not consulted or considered. Not only were the voters who sent these representatives from faraway states not interested in the District, but they were often openly hostile and opposed to their tax dollars being used for such a mundane purpose as improvement of the transportation network in a faraway city they already despised.

The effect on community and recreational uses of bridges in the District was substantial. The Capital Beltway was constructed to move commuters within the Maryland and Virginia suburbs surrounding the city, with spur highways funneling this high-speed traffic across the Potomac and into the city center on the unappealing Theodore Roosevelt and Fourteenth Street Bridges. The Anacostia Freeway along the eastern shore of the Anacostia River split the city in two, severing the neighborhoods east and south of the river from the communities and, more importantly, jobs in the center of the city. Spurs from the freeway at Benning Road, New York Avenue, East Capitol Street, Eleventh Street, and South Capitol Street crossed the Anacostia River on unattractive midcentury highway bridges.

These outrages finally ceased in 1972 when, after years of legal battles, the Federal Highway Administration abandoned its plans for the Three Sisters Bridge crossing the Potomac near the northern end of Roosevelt Island and into the Georgetown Waterfront. Although the destruction of neighborhoods to bring yet another major highway through the city to serve majority white suburbs was a major reason to object to the bridge, the environmental and visual effects would also have been devastating.

COMMUNITY-CENTRIC BRIDGES

In a 2006 article in *The Washingtonian*, Van Dyne states that the District's bridges "lack the scale and drama" of bridges in other cities. In his article, Van Dyne quoted Eric DeLony, retired chief of the National Park Service's Historic American Engineering Record, who described the District's bridges as "subtle and elegant and don't scream at you." These observations are generally true—the District has historically not had a widely known landmark bridge that serves as a symbol of the community.

Where landmark bridges are found, they are almost always near the top of the must-see lists for tourists. As noted in *Bridgespotting*, a stylized silhouette of a landmark bridge is often used to symbolize the community. However, in order to become an iconic symbol, a bridge must be prominently visible to the entire community, residents and visitors alike. This visibility is typically due to a bridge's height or central location. Until 2021, the District did not have a bridge with a highly visible superstructure for several reasons.

The first is that the business of Washington has always been more focused on government than industry. Although both Georgetown and Alexandria served as early ports, they never became as commercially important or as prominent as the shipbuilding or manufacturing centers of Baltimore, Philadelphia, New York, or Boston. The more important the port, the more likely that a bridge requiring a high superstructure, such as a prominent suspension bridge, will be constructed. Because the District did not have a major port, no early high-level suspension bridge was constructed. In fact, there are currently no suspension bridges in the District, whereas suspension bridges serve as prominent landmarks in San Francisco, New York City, Pittsburgh, and dozens of other cities.

A second reason a bridge never became a monumental symbol of the city is that such a symbol was effectively prohibited by law. In 1910, Congress passed the Building Heights Act, which restricted the height of all architecture in the city to a maximum of 130 feet. Also, the design of any new construction in the District is subject to review and approval by the U.S. Commission of Fine Arts. The purpose of these restrictions is to make sure that the only monumental symbols of the city are, in fact, the monuments. The Capitol dome and the Washington Monument are, effectively, the "official" architectural symbols of the District, and no other structures are allowed to displace them. No structures will ever be permitted that could interfere with the views of the Capitol dome, the Washington Monument, the Lincoln and Jefferson Memorials, and Arlington National Cemetery from as many locations in the city as possible. As all of these are prominently visible from the monumental core of the city along the Potomac River, it is likely that no bridge with a prominent superstructure will ever be allowed to cross the Potomac.

The lack of a landmark bridge in the District can also be attributed to the manner in which the city has been governed for most of its history. As discussed, it was not until the 1970s that the National Environmental Policy Act and Home Rule Act created the conditions that allowed local residents to participate in decisions regarding bridge construction.

This situation was not entirely without its benefits for the District's bridges. Neither Congress nor the appointed commissioners were particularly concerned about creating bridges to commemorate the city's political and cultural leaders, but this did not mean that attractive commemorative bridges were not constructed. By the 1960s, the Arlington Memorial Bridge had been constructed as a symbolic connection between Abraham Lincoln's North and Robert E. Lee's South, and three other bridges had been named for presidents: William Howard Taft, Woodrow Wilson, and Theodore Roosevelt. In the same time frame, the Francis Scott Key, John Philip Sousa, and Frederick Douglass Bridges were named for local residents who had an impact on national events or culture. However, these bridges commemorate on a national scale rather than a local scale.

In 1972, the cancellation of the Three Sisters Bridge represented the final death throes of the mid-twentieth-century idea that all transportation and traffic problems could be solved simply by constructing more and larger roads and bridges. Even before construction started on the Three Sisters Bridge, city planners had begun the move away from automobile-based solutions and toward mass transit. The first Metro stations opened in 1976, and the effort culminated with the opening of the Green Line into Anacostia in 1991.

With the movement away from automobile-based solutions, it appeared that the last major traffic bridges to be constructed in the District would be the 1964 Theodore Roosevelt Bridge and the 1972 addition to the Fourteenth Street Bridge complex, leaving the Potomac and Anacostia with a mix of mostly unattractive, utilitarian highway bridges. Key Bridge in Georgetown was at least historic and of a grand scale. Arlington Memorial was the only bridge that made any pretense at commemoration, but that was commemoration of national events as opposed to local or community values. A few of the Anacostia bridges, specifically Whitney Young at East Capitol Street, New York Avenue, and John Philip Sousa at Pennsylvania Avenue, were not unattractive, but not particularly special. But many of the bridges, including the old Frederick Douglass Bridge, Eleventh Street, Fourteenth Street, Woodrow Wilson, and Roosevelt Bridges were just unfortunate.

Fast-forward to about 2004, thirty years since a major bridge had been constructed in the city, when some of the older bridges needed replacing either because of structural deficiencies or because they had become a major hindrance to the transportation network. Two major political developments had occurred in the intervening years. The transportation network was now under the management of city officials who were accountable to the community through elections, and the federal agencies involved in bridge funding, construction, and permitting were now accountable to the community through the National Environmental Policy Act.

More important than the change in the political and governmental framework was the change in public attitudes toward nature, the environment, aesthetics, and outdoor recreation. These changes began in the 1960s and were part of the impetus for passage of the National Environmental Policy Act. They have only accelerated since then. It is not simply that the residents have demanded that community-friendly features be incorporated into their new bridges. The politicians, bureaucrats, engineers, and bridge designers want them as well. They live in the community, too, and understand how these features improve their own quality of life. It also gives them an enormous sense of pride and job satisfaction to know that they are doing far more than just moving traffic. They are also creating civic improvements that will enhance the community for decades to come.

Since 2004, this has been put into practice on the replacement of four major bridges: Ethel Kennedy Bridge at Benning Road in 2004; Woodrow Wilson in 2007; Eleventh Street between 2011 and 2013; and Frederick Douglass in 2021. Each of these bridges has wide sidewalks connecting to the extensive bike trail system, providing pedestrian access from residential neighborhoods to Heritage and Kingman Islands, the Washington Nationals baseball stadium, and the convention and casino developments at National Harbor.

These changes were subtle at first, mostly noticeable only to cyclists and pedestrians, and not highly visible to drivers and passengers in cars. However, this subtlety was blown out of the water in 2021 as the high white arches of the new Frederick Douglass Memorial Bridge began to rise above the Anacostia. The Frederick Douglass Bridge became the first District bridge with a superstructure of any substantial height, and it is in one of the most visible locations in the entire city, surrounded by glittering new office towers and condominiums. If residents had not yet noticed the subtle changes implemented at the Ethel Kennedy, Eleventh Street, or Woodrow Wilson Bridges, they noticed them now.

Finally, even with the limitations imposed by the Building Heights Act, the city has begun to use bridge superstructures and decorative lighting to turn its bridges into public artworks. The Frederick Douglass and Yards Park Bridges are both attractive public sculptures by day and spectacular additions to the nighttime cityscape. Even the utilitarian midcentury Francis Case Bridge comes alive at night with attractive lighting on its stone-faced piers.

And that is just the start. Plans are in place for several more bridge renovations, replacements, and even new pedestrian bridges. Bridge construction in the District will never be the same, which is a good thing for tourism, recreation, traffic, and the interconnection of communities across the city. Any future bridges constructed on either the Potomac or Anacostia will be a collaborative effort involving both the District Department of Transportation (DDOT) and one or more federal agencies. DDOT is responsible to the city's elected mayor and city council and, therefore, the voters. Similarly, all future bridges will require compliance with the public participation components of the National Environmental Policy Act. Never again will there be a bridge constructed in the city without full participation of its residents, and residents everywhere have made clear that they demand pedestrian access, connections to parks and trails, decorations, and recreational features.

3

POTOMAC RIVER

There are two vastly different Potomac Rivers within the boundaries of the District. From the Maryland state line to Georgetown, fresh water flows over crystalline bedrock through the Potomac Gorge. The Potomac Gorge is a narrow valley eroded by the Potomac River into the bedrock of the Piedmont. Fifteen miles long and lined with bedrock cliffs known as the Potomac Palisades, the Potomac Gorge is the location of Great Falls and Little Falls. Within the Potomac Gorge, the river's elevation drops from about 140 feet to ten feet above sea level.

The shores of the Potomac within the Potomac Gorge consist almost entirely of wooded parkland. On the Maryland side, the C&O Canal and the adjacent Canal Road run parallel to the river, within a few hundred feet of the shore. The canal and Canal Road are squeezed in between the river on the west and the bluffs of the Potomac Palisades on the east, and both sit at an elevation only slightly higher than the river itself. Both are located within the C&O Canal National Historical Park, operated by the National Park Service.

On the Virginia side, the George Washington Memorial Parkway, operated as a unit of the National Park Service, is parallel to the shore. However, the parkway generally follows the top of the Palisades' bluff, which is significantly higher than the elevation of the river. Because this elevated position offers stunning views of the river, the parkway is a popular highway for visitors approaching the District from the north and west.

The shoreline of the Potomac Gorge was never developed as docks for shipping because it is adjacent to the freshwater, free-flowing part of the Potomac River on the Piedmont bedrock. It also never underwent substantial urban development because the slopes of the bluffs on either side of the river were too steep. Therefore, there was little need for bridges across the Potomac Gorge. In more than 230 years since the city's founding, only one bridge, the one at Little Falls now known as Chain Bridge, has been constructed across the gorge within the city boundaries.

Below Little Falls, the river leaves the Potomac Gorge, reaches sea level, and becomes a tidal estuary near Georgetown. The remainder of the eastern shore of the Potomac south of Georgetown was originally tidal marshes and swamps until the entire shoreline was reengineered in the late nineteenth and early twentieth centuries. The river was dredged to deepen the main channel of the Potomac and to create a harbor, which is today the Washington Channel. The dredged material was used to fill in swampland south of the White House, creating the western end of what is today the National Mall, home to the Washington Monument, Lincoln Memorial, and numerous other museums and memorials. The material was also used to create the Tidal Basin, known to residents and visitors as the location of the Jefferson, Franklin Roosevelt, and Martin Luther King Jr. Memorials, all surrounded in spring by an explosion of pink cherry blossoms.

The western shore of the Potomac south of Little Falls underwent similar reengineering and urban development. In this area, dredging and deposition of dredged materials were used to convert tidal marsh to developable land for Columbia Island, the Pentagon, and Reagan National Airport.

The flat land on both shores of this lower part of the Potomac made urban development and associated highway and bridge construction possible. This is where the bridge crossings of the Potomac River are focused.

CHAIN BRIDGE

Chain Bridge crosses the C&O Canal and Potomac River within the Potomac Gorge, approximately three miles northwest of Georgetown.

Because of its setting within the gorge, the bridge is not particularly well-known to District residents or to those in the Maryland and Virginia suburbs. The bridge carries a narrow, winding two-lane roadway and serves mostly to funnel commuting traffic from the close-in Virginia suburbs to

Canal Road and into Georgetown. However, the short roadway ends in a *T* at the eastern end of the bridge, so it generally carries local traffic as opposed to through traffic.

While the setting of the bridge in the gorge and crossing Little Falls is scenic, the bridge itself is not decorated, attractive, or unusual in a way that would interest observers passing by in cars or attract tourists. Constructed in 1938, the bridge appears to be one of thousands of relatively unappealing, utilitarian, midcentury steel-plate girder bridges. However, its plain appearance belies the historical importance of the location of the bridge and its role in the history of Georgetown and the District.

The location of Chain Bridge was the first bridge crossing the Potomac in 1797, but the importance of the site goes back even further. At this location, a small stream, Pimmit Run, breaks through the Palisades on its way to the Potomac, forming a narrow gap in the cliffs on the Virginia side of the river. This narrow gap became the focus of early Native American trails, making this location convenient for river crossing. Once Europeans arrived, the established river crossing and the flowing water of Pimmit Run made it a suitable location for a gristmill as early as 1719. An industrial settlement developed and, by the 1790s, included a distillery, brewery, granary, cooper shop, blacksmith shop, wool factory, cloth mill, and paper mill.

The river crossing also served a purpose for the port of Georgetown, one of many incidents in what became an ongoing rivalry between Georgetown and the nearby port of Alexandria, each vying to dominate trade in the region. The main agricultural and industrial settlements in the northern part of Virginia were in Leesburg, located on the western bank of the Potomac about thirty miles northwest of Georgetown. In 1791, business interests in Georgetown formed the Georgetown Bridge Company, with the purpose of developing a bridge at the mouth of Pimmit Run. The bridge would complete a Georgetown-Leesburg Road, encouraging commercial traffic to bypass Alexandria by crossing the river directly into Georgetown.

The first of eight generations of bridges at the location was completed in 1797. It was called the Falls Bridge according to some sources and the Potomak Bridge according to others. It opened in October 1797 with tolls of three cents for pedestrians, eight cents for man and horse, twenty-five cents for a two-wheel cart, and fifty cents for a four-wheel cart.

This first bridge was a wooden covered bridge, which collapsed in 1804. One source, *Bridges and the City of Washington* by Donald Beekman Myer (1974), states that the bridge collapsed due to rot, but a more colorful story is offered in another source, *Washington's Bridges: Historic and Modern*, by the

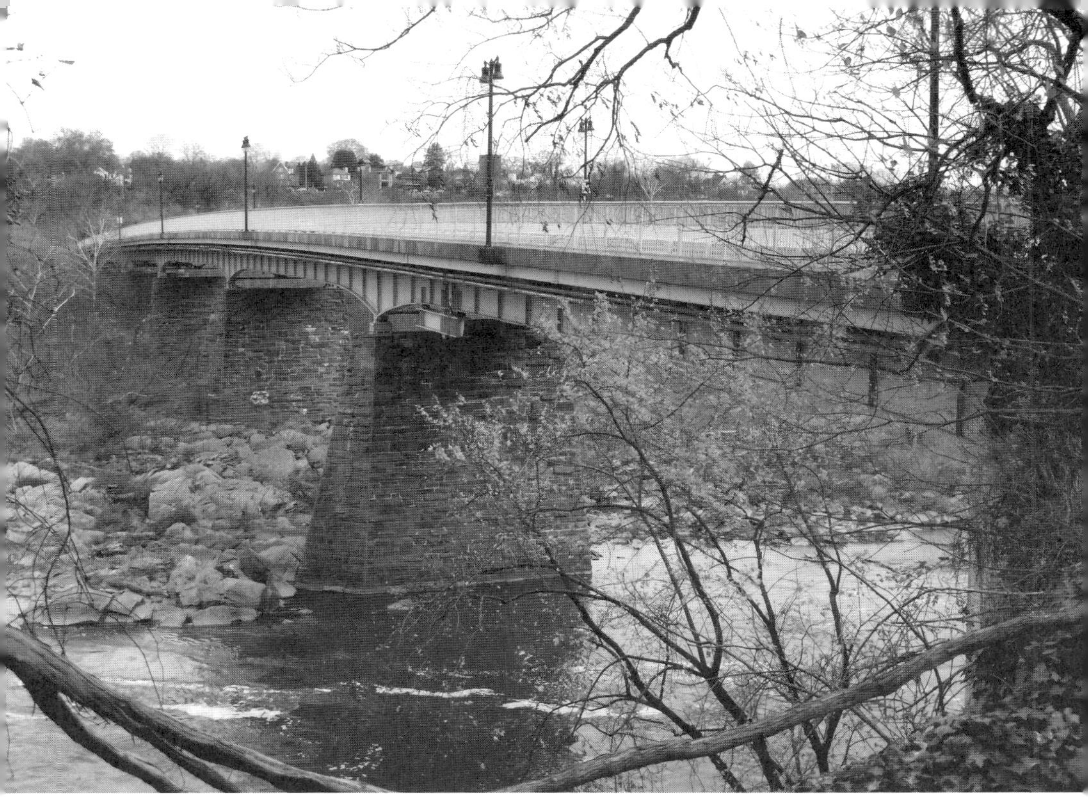

View of Chain Bridge at Little Falls, looking toward homes atop the bluff of the Potomac Palisades.

District Department of Highways (1956). The bridge was used to bring cattle from agricultural areas to auction in Georgetown. Because cattle are sold by weight, the owners would stop at the mouth of Pimmit Run to allow the cattle to drink water, making them heavier just before they arrived at market. According to this source, the weight of these waterlogged cattle was too much for the bridge, leading to its collapse.

The bridge was reconstructed, again made of wood, but that second bridge either collapsed in a flood (according to Myer and Bryan) or burned (according to the District Department of Highways) only six months later. Construction of a third bridge began in 1808 and was completed in 1810. This bridge was a wooden truss, suspended by chains from a stone tower on each shore, giving the name "Chain Bridge" to the road and to all future bridges at the location. The third bridge was 136 feet long and fifteen feet wide, and it was destroyed by a flood in 1812.

The fourth bridge, a similar suspension bridge, was constructed and later damaged by a flood in 1815. Although its lifespan was short, this fourth bridge played an important role in American history. As the British approached the District in August 1814, Secretary of State James Monroe directed clerk Stephen Pleasonton to remove the Declaration of Independence and other important documents to a safe location. Because the British were approaching from the east, Pleasonton had the documents packed in a cart and transported to the west, where he temporarily stored them in the mill at the Virginia end of the fourth bridge.

The fifth bridge was constructed in 1840. It also eventually succumbed, so a sixth bridge was constructed in 1852, but this time the bridge was constructed on a series of stone piers instead of being suspended on chains. A flood in 1870 destroyed it, but the piers remained intact. A seventh bridge with eight iron-truss spans was constructed on these piers in 1874. In 1926, erosion was found under the abutment on the Virginia shore. Speed and weight limits were instituted, and then the bridge was closed to traffic and remained open only for pedestrians. Meanwhile, the type of traffic using the 1874 bridge had changed, and the twenty-foot width made the bridge obsolete. Following another flood in 1937, the current bridge was constructed, using steel girders but still sitting on the stone piers from the 1850s.

Today, this history is presented on several historic information placards at both ends of the bridge. The bridge can be accessed by parking in a small lot along Canal Road on the eastern end or in a small lot serving as a trailhead at the end of Glebe Road on the western end. In addition to the bridge's history, the placards discuss the history of the C&O Canal (which is crossed by the bridge) and the industrial developments at the mouth of Pimmit Run.

Because it is squeezed in between cliffs of the Palisades, there is no commercial or residential development directly on the ends of the bridge. Even though thousands drive by the end of the bridge where it intersects Canal Road every day, the setting seems relatively remote and rustic, with both banks of the Potomac lined with woods and bedrock cliffs. The bridge has a sidewalk on its northern side facing Little Falls, making it a good place to view the Potomac Gorge.

POTOMAC AQUEDUCT

The original plan for the C&O Canal, as conceived by George Washington, was to create a waterway connection between Georgetown and the Ohio River basin. In 1830, even before the canal was completed, Congress chartered the Alexandria Canal Company to construct and operate an extension across the Potomac and into Alexandria. The extension consisted of an aqueduct, which was a watertight trough sitting on piers, allowing canal barges to float high above the Potomac while still allowing shipping to occur on the river beneath the aqueduct. The western end of the structure, which was variously designated with the names "Potomac Aqueduct," "Aqueduct Bridge," or "Alexandria Aqueduct," was then connected to Alexandria with a seven-mile-long canal in Virginia. The aqueduct and Alexandria Canal allowed canal traffic from western Maryland to bypass Georgetown and instead use port facilities in Alexandria.

Construction of the Potomac Aqueduct began in 1833, and it opened in 1843. It served as an important crossing of the Potomac River for eighty years, until 1923. Although it no longer exists, its history provides important context for other bridge-related developments in the history of the city, including the ongoing rivalry between Alexandria and Georgetown for primacy as the industrial and commercial port for the District.

At the onset of the Civil War in 1861, the federal government took over the aqueduct from its private owner. The trough was drained, and a roadway was installed in the bottom. The connection of the trough to the C&O Canal was severed, and the new roadway was connected to what is now M Street in Georgetown. The new bridge was used during the war to carry Union army supply wagons from the District into northern Virginia.

In 1866, the bridge was returned to the owners, who leased it to the Alexandria Railroad and Bridge Company. They removed the old trough and constructed a new one, still of wood. Then, they put arched trusses on top of the trough and added a wooden toll road over the arches.

Residents and commercial interests complained about high tolls, and massive leaks from the trough interfered with boat traffic below. In 1886, the federal government purchased the bridge, removed the trough, and constructed a series of iron trusses across the original stone piers, ending the use of the structure as an aqueduct. The new bridge, freed from tolls, opened on April 11, 1888, and the opening was marked by large celebrations in Georgetown.

In 1890, it was reported that the bridge was too narrow and weak for the traffic and some of the piers were deteriorating. A replacement bridge was considered as early as 1902, and the design of the current bridge was initiated in 1916. This bridge, named after Francis Scott Key, opened in 1923. At that time, the ends of the Potomac Aqueduct were cut off, but it remained in place for an additional ten years. Eventually, the superstructure was removed in 1933 by the New Deal–era Civil Works Administration. In

1962, at the direction of the U.S. Commission of Fine Arts, all but one of the stone piers was removed from the river.

Although the Potomac Aqueduct was removed in 1933, it lives on in ways that can still be enjoyed by history buffs and bridge enthusiasts. The lifetime of the bridge from 1843 to 1923 coincided with the invention of early photography and airplanes. Historic aerial photos of the Potomac Aqueduct, with the picturesque spires of Georgetown University in the background, are available. For those residents who are familiar with the current views of the university and Key Bridge, these black-and-white photos of the industrial-looking Potomac Aqueduct occupying the same space can be surprising.

The second way the Potomac Aqueduct can still be appreciated is by visiting the northern abutment, known as the Aqueduct Arch, which remains in place just a few feet north of Key Bridge. The arch is the terminus of the Capital Crescent Trail, which runs parallel to the C&O Canal for three miles west of Georgetown and then passes through the arch. The trail passes a few buildings housing boating clubs and ends in a parking lot along Water Street beneath the on-ramps of the Whitehurst Freeway.

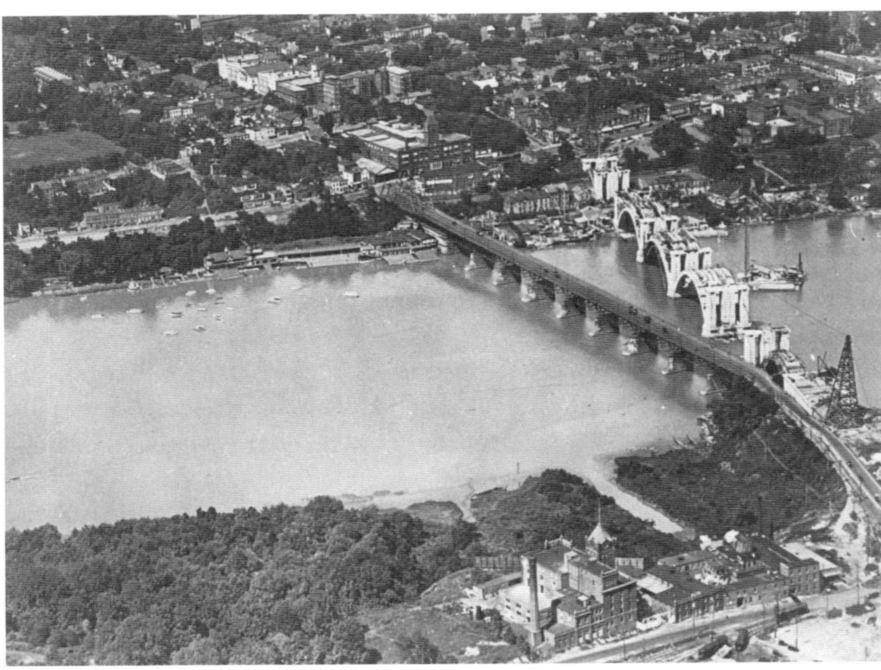

A 1922 aerial view of Key Bridge (under construction) and Potomac Aqueduct, looking toward Georgetown University. *Author's collection.*

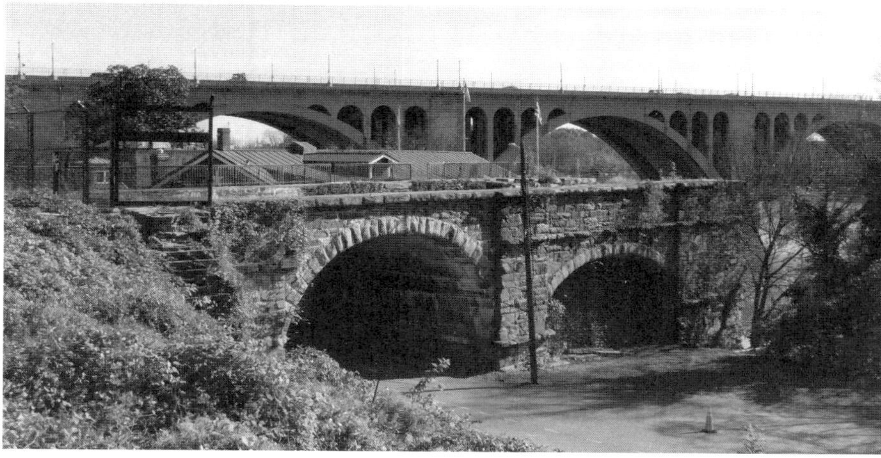

View of Potomac Aqueduct Arch in Georgetown, with Key Bridge in background.

Anyone interested in the District's transportation history should visit this remnant of the Potomac Aqueduct. The location was the convergence of the southern terminus of the C&O Canal and the Georgetown Branch of the Baltimore and Ohio (B&O) Railroad, as well as the southern end of the Foundry Branch Trolley Line, the northern ends of the Potomac Aqueduct and Key Bridge, and the western end of the Whitehurst Freeway. All of these different pieces of industrial and transportation infrastructure come together at the Aqueduct Arch, just a few feet north of Key Bridge.

The arch looks old, as it is constructed of beige stone blocks that have been blackened by time and weather. It is about forty feet long and twenty feet wide and is topped by a security fence to stop people from walking on top. No placards or other identifiers are there to inform passing hikers and cyclists of the historic importance and unusual function of the arch.

Another extant remnant is a stone pier in the river on the Virginia end of Key Bridge. This pier is briefly visible to passing traffic on northbound George Washington Memorial Parkway as the traffic emerges from beneath Key Bridge. It can be reached on foot by following the Potomac Heritage Trail north from the parking lot at Theodore Roosevelt Island. The trail is not paved, so it is not easily accessible by bicycle and requires a substantial hike on foot from the parking lot. The abandoned stone pier is only about fifty feet from the shoreline, almost directly under Key Bridge. From the shoreline, you can see the remnant of the Arch on the Georgetown shore and envision the canalboats floating high above the Potomac nearly 200 years ago.

FRANCIS SCOTT KEY BRIDGE

The Francis Scott Key Bridge, constructed in 1923 as a replacement for the aging Potomac Aqueduct, is the oldest bridge over the Potomac. The bridge crosses the Potomac River directly in the center of Georgetown, connecting M Street to the skyscrapers of downtown Rosslyn, Virginia. As Rosslyn's development boomed over the past thirty years, the bridge has played an increasingly important role as a pedestrian commuting link between Rosslyn's business center and Georgetown University, with its surrounding residential areas. The bridge was added to the National Register of Historic Places in 1996, one of only seven District bridges so recognized.

The bridge is a large-scale reinforced-concrete arch bridge. It was originally seven spans long and had an eighth span added on the Virginia end in 1939 to accommodate construction of the George Washington Memorial Parkway. In addition to the Potomac River and the parkway, the arches cross the C&O Canal and Whitehurst Freeway in Georgetown. The total length is 1,781 feet, and the original deck was sixty-six feet wide, with two sixteen-foot traffic lanes, two eight-foot-wide sidewalks, and a center lane for trolleys. The trolleys were eliminated in 1955 with the widening of the roadway. At the same time, concrete parapets were removed and replaced with a cast aluminum railing. In 1987, a new deck, ninety feet wide with expanded sidewalks, was constructed, helping improve pedestrian and bicycle access between Georgetown and Rosslyn.

Key Bridge is one of the more visually prominent bridges in the District. In a city where the ability to construct superstructures on bridges is limited by the Building Heights Act of 1910, only a few concrete arch bridges that

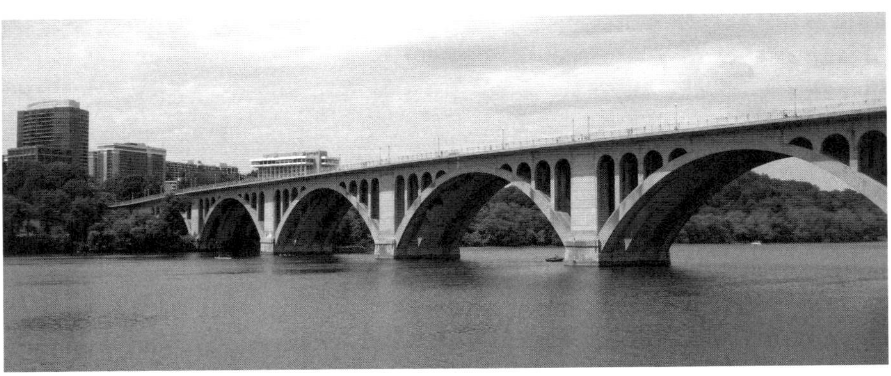

View of Francis Scott Key Bridge taken from the Georgetown Waterfront.

connect bluffs on either side of an incised valley can be high enough to be visible from a distance.

Located at the downstream end of the Potomac Gorge, Key Bridge is one of those bridges. South and east of Georgetown, the land on either bank of the Potomac is flat, with an elevation only slightly higher than sea level. However, north and west of Georgetown, the river flows through the Potomac Gorge and is lined with high bedrock bluffs on either side. This is one of the reasons the spires at Georgetown University are so picturesque—they are not only high, but the buildings also sit on top of a hill overlooking the river. To connect to the roadway system on higher areas on either side of the river, Key Bridge consists of high arches, with a deck situated about seventy-two feet above the river.

The arches of the bridge are not only high but decorative as well. It is an open-spandrel arch bridge, with the sides of the arches punctuated by a graduated series of secondary arches. The central arch is the widest at 208 feet, flanked by two arches at 204 feet, and these are flanked by two arches at 187 feet. Each arch is a tripartite structure consisting of three adjacent steel arches set in concrete foundations. The construction was done with a concrete plant on shore and another floated into position and then sunk to rest on bedrock. The steel ribs of the arches were floated into place and then encased in concrete.

The bridge was originally designed to be a two-level bridge, but the design was simplified due to financial limitations during the First World War. It is not related to the L'Enfant Plan and does not align with any of the District's monuments or avenues. The original plans called for removing buildings along the northern side of M Street in Georgetown to construct a monumental park linking the bridge to Rock Creek, but this was not done.

The bridge was designed in the Classical Revival style by architect Nathan C. Wyeth, who also designed the West Wing of the White House and numerous other structures in Washington. The design of public structures in the District must be approved by the U.S. Commission of Fine Arts, which was established in 1910. In 1916, at the time the bridge was designed, the City Beautiful Movement was in full force, and the commission was favoring high-profile, decorative concrete arch bridges, such as 1907 Taft Bridge.

The name of the bridge derives from Francis Scott Key, the author of "The Star-Spangled Banner." Key lived in a house just west of the Georgetown end of the bridge from 1803 to 1833. During the British invasion of Washington in 1814, Key, who was an attorney, went to the Royal Navy flagship to negotiate the release of prisoner Dr. Beanes. The negotiation succeeded, but Key was detained to ensure that he could not reveal British plans to attack Baltimore. He remained on the ship and witnessed the attack on Fort McHenry. When he saw the American flag still flying over the fort in the morning, he penned the poem that became the national anthem on the back of a letter.

The dedication to Key is described on a bronze plaque near the Georgetown end of the bridge. The plaque was erected by the National Society of United States Daughters of 1812 on April 21, 1924. The Key house was still in place when the bridge was constructed, but it was demolished in 1947. A small park at the northeastern corner of the bridge, where it intersects with M Street, is named for Key and has an informational placard discussing these events.

The view from the sidewalk is one pleasant attraction of a walk along Key Bridge. Georgetown University towers above the eastern end of the bridge, while Rosslyn's modernistic skyscrapers tower above the western end. In between, the scenery along the river includes trees lining the Georgetown Waterfront on the north and trees of Theodore Roosevelt Island on the south. Three buildings are prominently visible along the river just past Roosevelt Island: the Kennedy Center for the Performing Arts, the infamous Watergate hotel and apartment complex, and, peeking out above the Watergate, the spire of the Washington Monument.

THEODORE ROOSEVELT BRIDGE

The transformation of the United States into an automobile-centric society following World War II provided a massive improvement in the standard of living for huge numbers of people. For the first time, workers could benefit from the concentration of industry and commerce in urban centers during the day but then enjoy the pleasures of a single-family home, fenced yard, and open spaces at night and on weekends. All that was needed was an automobile and a roadway connecting the two locations.

The idea was so brilliant that millions of Americans wanted to take part, leading to an explosion in suburban development and, with it, the need to build highways and bridges to accommodate commuters. The pressure to build and continually expand this infrastructure was so intense that communities across America completely lost sight of aesthetics.

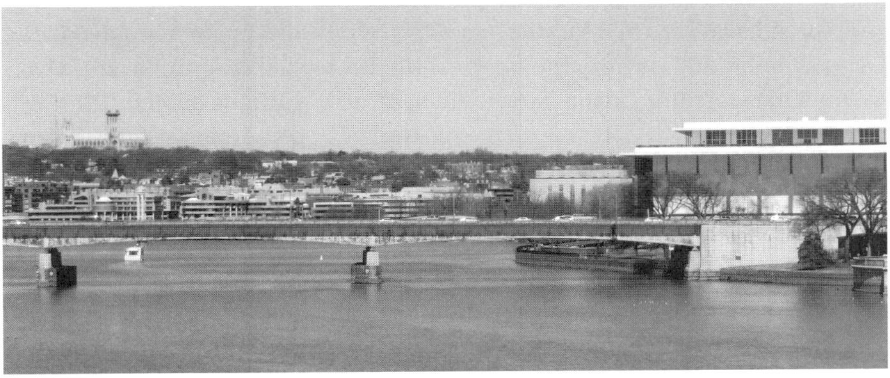

View from the sidewalk of Arlington Memorial Bridge, including the Washington Cathedral, Kennedy Center, and Theodore Roosevelt Bridge.

There are few better examples of the long-term consequences of this than the Theodore Roosevelt Bridge. With respect to tourism and aesthetics, the bridge is almost certainly the most disappointing in the District. Donald Beekman Myer's 1974 survey of Washington's bridges refers to the bridge as the city's "most criticized."

The Roosevelt Bridge carries commuting traffic from northern Virginia into the District on Interstate 66, one of the busiest highways in the region. It is an extremely unattractive steel-plate girder highway bridge, which would be unfortunate in any location. However, it is not in a random location; it is in the monumental core of the nation's capital.

The bridge, which opened in 1964, has few, if any, redeeming qualities. Neither the bridge nor its location has any importance in the history of the city or the nation. The bridge was not constructed with innovative engineering and does not display any decorations. There is a dedication plaque to Theodore Roosevelt, but it is located on the trail that passes beneath the bridge, in a spot unlikely to be seen by drivers or anyone using the sidewalk to cross the bridge.

A generous assessment would point out that the bridge is on the northern edge of the monumental core, so it only mars the viewscape for a fraction of visitors. The bridge is also low in profile, with no superstructure, thus reducing its visibility even further. Unusual among interstate highway bridges, it does have a narrow sidewalk, which provides some support to the city's recreational trail system. On the Virginia end of the bridge, the sidewalk connects to a ramp leading down to the Mount Vernon Trail, a short distance south of the Roosevelt Island parking lot. On the District end, the sidewalk connects to the street-level sidewalk system at the Kennedy Center. As of this writing, the deck on the bridge is being replaced, which will include widening of the narrow sidewalk to a width of ten feet. This will substantially improve the use of the sidewalk for pedestian and bike access to Roosevelt Island and the Mount Vernon Trail.

The most unfortunate feature of the bridge is its effect on the monumental viewscape of this part of the Potomac waterfront. In this stretch of river, only a little more than a mile in length, tourists and residents will find all of the most iconic attractions and memorials of the nation's capital, including the sparkling white Kennedy Center for the Performing Arts, Arlington Memorial Bridge, the Lincoln Memorial, and Arlington Cemetery. The entire Potomac shoreline explodes with pink cherry blossoms every spring, attracting hundreds of thousands of tourists and residents to downtown. The unattractive Roosevelt Bridge is visible from almost anywhere within this area and forms a major part of the view from Arlington Memorial Bridge, effectively blocking the view of the Kennedy Center. Making things worse, this impact was well known even before the bridge was constructed. In his summary of the "Bridges Across the Potomac" in the *Records of the Columbia Historical Society* in 1956, less than ten years before the bridge opened, Robert C. Horne relates that the National Park Service was opposed to a bridge at this location because of its interference with views of the monuments, favoring a tunnel instead.

ARLINGTON MEMORIAL BRIDGE

Perhaps the bridge best known to District tourists and visitors, and the one bridge that may serve as a recognizable symbol of the city, is Arlington Memorial Bridge.

The District's monumental core is centered on the Mall, which forms an open vista extending almost three miles from the Capitol on the east, past the National Gallery of Art and museums of the Smithsonian, to the Washington Monument and farther west to the Lincoln Memorial on the shore of the Potomac. At the river, this core occupies about a mile of the eastern shoreline. The northern end of the core is marked by the Kennedy Center, and the southern end is the Tidal Basin and Jefferson Memorial. On the top of the bluff directly opposite this core, continuing the white marble theme of the monuments and museums, are the Custis-Lee Mansion and the graves of Arlington National Cemetery. This corridor is the most recognizable part of Washington, where the nation's founders and heroes are celebrated and where the wars fought to give the citizens their freedoms are commemorated. Within this core, serving as a literal and symbolic link between the nation's most important cemetery and the monuments to its heroes, is one bridge: Arlington Memorial.

The roadway carried by Arlington Memorial Bridge is more commemorative than functional. It is not an interstate highway, nor does it connect to one. With two lanes in each direction, it certainly carries its share of commuters, but the bridge is not situated in a location that makes it useful for commuting. The roadway carried by the bridge is not a through road, so neither end continues directly into a business or residential district.

Instead, the bridge is an integral part of the complex of memorials, parks, traffic circles, parkways, and displays of public art that comprise the city's tourist center. With its carved white-granite balustrades and benches, sparkling-white color, and monumental-scale statues at each entrance, it is integrated into the theme of the surrounding memorial and ceremonial structures, making it an important memorial on its own.

On the eastern end, the roadway merges into the circle surrounding the Lincoln Memorial, and the roadways off this circle lead to the Mall's memorials, monuments, and museums. The western end of the bridge is on Columbia Island, a narrow park that is partially forested and landscaped but is mostly occupied by a complex network of on-ramps and off-ramps for the George Washington Memorial Parkway. The roadway on the bridge merges into another traffic circle on Columbia Island, which, in addition to off-ramps for the parkway, connects to the entrance road for Arlington National Cemetery. Overall, the bridge leads only to monuments, memorials, and parks on either end and is not particularly helpful to commuters, which reinforces the bridge's commemorative function.

The development of the Custis-Lee Mansion and surrounding property into a national cemetery took place during the Civil War. Congress originally proposed construction of a bridge at this location in 1887 to serve as a memorial to Ulysses S. Grant, but it was not constructed at that time. In 1887, what is today known as the Mall did not exist. The area south of the White House was swampland until about 1900, when the Army Corps of Engineers began filling in the tidal flats west of the Washington Monument. At the same time, the Office of Buildings and Grounds of the War Department, which was responsible for the Mall and the locations of public buildings, developed a plan that included a bridge connecting the Mall to Arlington. By 1902, the McMillan Commission had developed a plan, based on the City Beautiful Movement, to develop boulevards and monuments that matched the grandeur of those in London, Paris, Rome, Venice, and Vienna. The models and plans were presented to the Senate and then to the public at the Corcoran Gallery on January 15, 1902. The plan included development of Potomac Park around the Tidal Basin and a bridge connecting the Lincoln Memorial at the western end of the Mall to Arlington Cemetery.

Still, the bridge was not constructed immediately. Debate continued on the various designs, orientations, heights, and commemorative functions of

View of Arlington Memorial Bridge and Lincoln Memorial from the Mount Vernon Bike Trail.

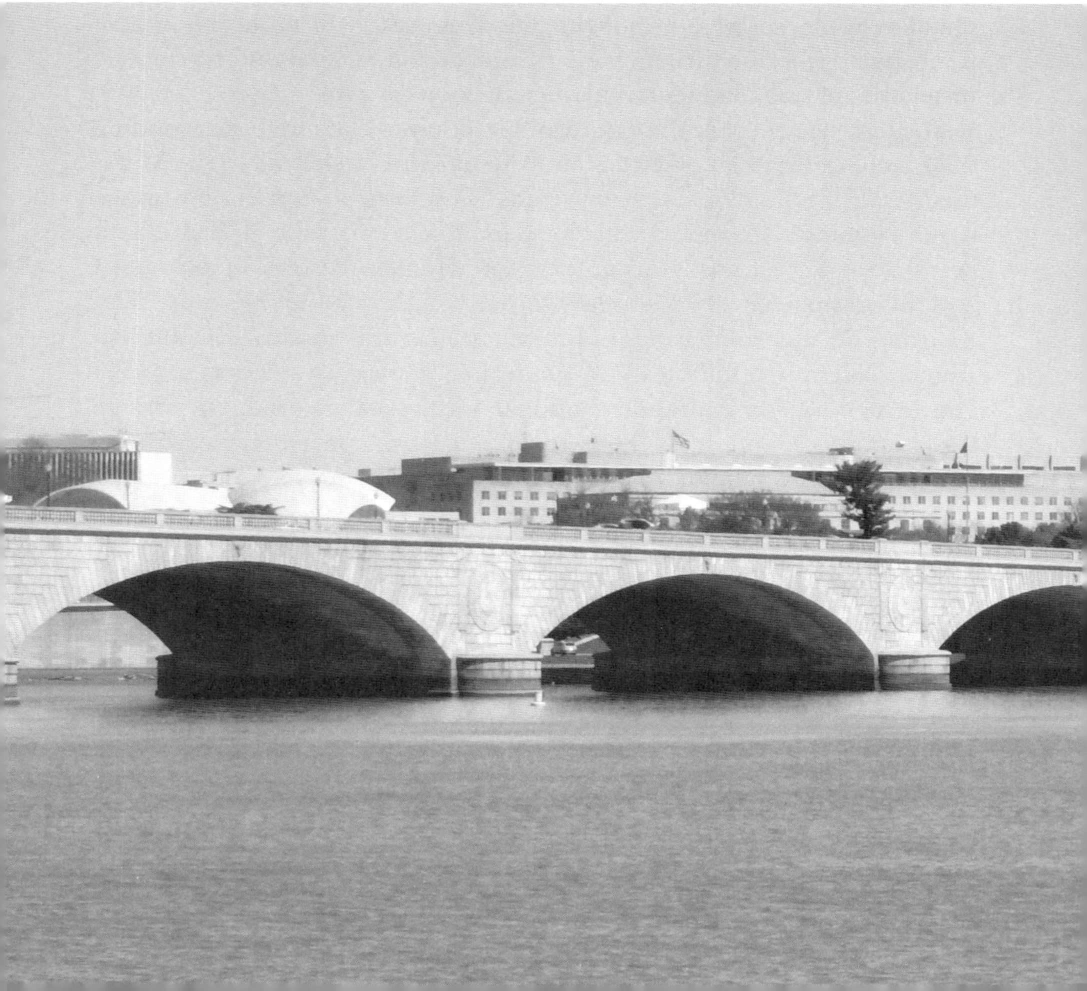

the bridge. Finally, President Warren G. Harding, caught in a three-hour traffic jam trying to reach the dedication of the Tomb of the Unknown Soldier at the cemetery in 1922, secured funding for a new commission to study the bridge. Construction was authorized in 1925, and the bridge opened to traffic in 1932. The bridge was listed on the National Register of Historic Places in 1979.

The bridge consists of a central steel arch and four concrete arches on each side. The concrete arches are enclosed within a white granite façade that complements the marble and granite memorials on the National Mall and the cemetery. The white granite balustrade, tops of the piers along the sidewalk, and flanking granite benches are carved with geometric designs. The closed spandrels of the bridge are decorated with large carved medallions depicting eagles. The bridge is only about twenty feet high so as to not obstruct any views of the monuments.

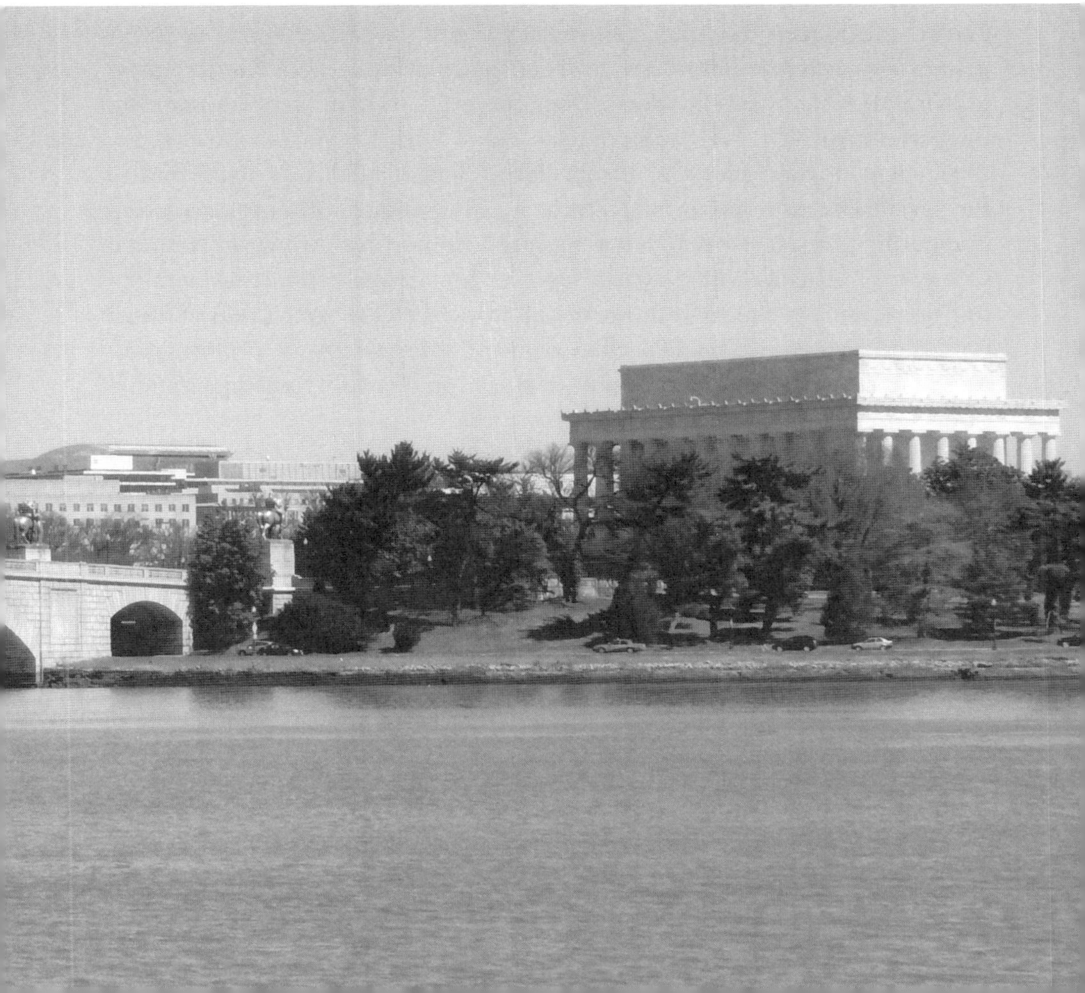

On the original bridge, the central arch was a drawspan, which allowed commercial shipping to reach Georgetown. However, later bridges on the Potomac blocked off this access by the early 1960s, and the drawspan had not been opened in more than fifty years. In 2015, corrosion of the steel in the bascule span required closure of the outer lanes and parts of the sidewalks. Temporary repairs allowed these to reopen, but with weight limits.

The bridge was fully renovated and rehabilitated in 2022. Because the drawspan was no longer functional, it was removed and replaced with a fixed steel span. However, to maintain as much historic integrity as possible, the original bascule machinery and operating rooms, hidden within the pier below the deck of the bridge, were preserved. The location of the operations room is identifiable by a set of white granite stairs leading down from the northern sidewalk to a bronze door.

The original steel façade that formed the exterior of the bascule span was also removed, repaired, and then replaced on the new fixed span in 2022. Also, the carved granite balustrades, consisting of more than five thousand individual pieces of granite, were dismantled, taken off-site for cleaning, and then reconstructed.

When it opened in 1932, the bridge was dedicated as a memorial to the sacrifices of the nation's veterans, a fitting dedication considering the manner in which it serves as a ceremonial entrance into Arlington National Cemetery. It was also intended to serve as a symbolic connection between the North and the South in the years following the Civil War, and the direct link from Abraham Lincoln's Memorial to the former home of Robert E. Lee across the river cannot be ignored. However, there are no plaques on the bridge discussing a specific memorial purpose.

Instead, gigantic allegorical monuments flank both ends of the bridge, as well as the nearby entrance to the Rock Creek and Potomac Parkway. On the eastern end, the monuments were a gift from the people of Italy to the United States in 1950, and the bronze statues atop the white granite bases were cast in Florence. The two bronzes on the eastern end of the bridge are equestrian statues representing *Sacrifice* and *Valor* as *The Arts of War*. The two Pegasus bronzes on the nearby entrance to Rock Creek and Potomac Parkway are *Aspiration* and *Literature*, and *Music* and *Harvest* representing *The Arts of Peace*.

On the western end of the bridge on Columbia Island are four twenty-five-foot-high granite monuments topped by stone-carved eagles. Two of the monuments flank the western entrance to the Arlington Memorial Bridge, and two others flank the entrance to the Boundary Channel Bridge, which is a short extension of the main bridge across a narrow channel separating Columbia Island from the river's western shore in Virginia. The construction type and appearance of the Boundary Channel Bridge are the same as the main bridge, creating a single uniform crossing from the Lincoln Memorial to the cemetery.

Another meaningful ceremonial function of the bridge is that it provides a direct connection between the U.S. Capitol and Arlington National Cemetery, completing the final route taken for funeral processions of individuals who have lain in state in the Capitol Rotunda. Three historic information placards are displayed on the southern sidewalk. Two of these describe the rehabilitation project and include an interesting photograph of National Park Service staff inspecting the interior of the operations room prior to the rehabilitation. However, the third describes the ceremonial role of the bridge in funeral processions of prominent national leaders, with a photo of John F. Kennedy's funeral cortege crossing the bridge in 1963.

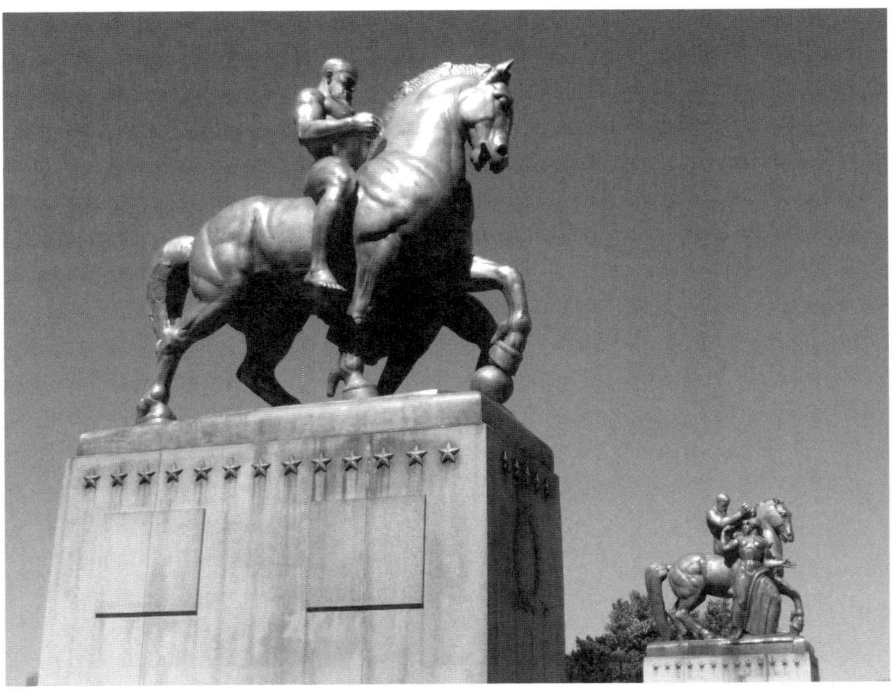

Allegorical bronze statues greeting drivers onto Arlington Memorial Bridge.

Accessing the bridge by foot or on bike is more complicated than most of the other District bridges. The bridge's sidewalk can be reached using sidewalks on surface streets, but it does not connect directly to the regional hiking and biking trail system on either end. Although the seventeen-mile-long Mount Vernon Trail passes directly under one of the bridge's arches on the western end, there is no easy access trail up to the bridge's sidewalk. It can be done but requires crossing active traffic lanes on the George Washington Memorial Parkway.

This difficult access to the bridge for tourists and visitors is a shame, because the bridge provides some of the best views of monuments and other important architecture in the city. Because the terrain is flat and Washington has no tall buildings, the views do not include impressive landscapes or city skylines east of the river. However, the view does include almost every well-known monument in the city. The view to the east, both from the sidewalk and the roadway, is dominated by the Lincoln Memorial, which sits at the end of the bridge. Just past the Lincoln Memorial, the Washington Monument looms over the trees. Depending on the time of year and whether the trees are filled out, there are also views of the Capitol dome and the Jefferson Memorial. In the spring, this view is lined with blazing-pink cherry tree blossoms.

To the west, the view is dominated by the Custis-Lee Mansion, also known as Arlington House, which was the former home of Robert E. Lee. The mansion was appropriated by the Union during the Civil War and still serves as the centerpiece of Arlington National Cemetery. The mansion is surrounded by rows of white gravestones on the slope of the hill overlooking

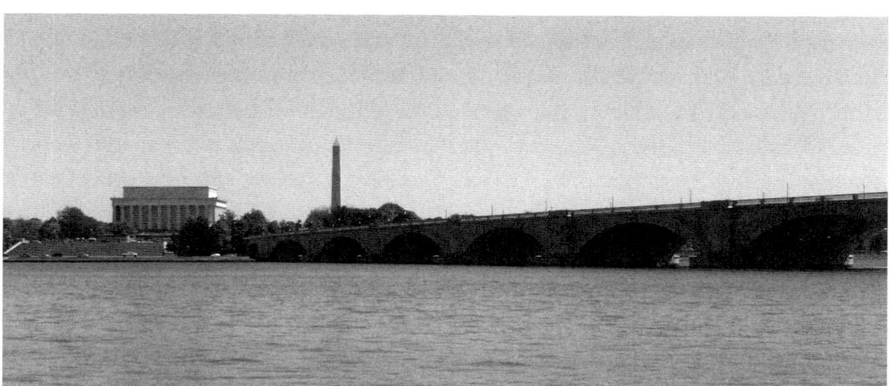

View of Arlington Memorial Bridge, Lincoln Memorial, and Washington Monument, taken from the Mount Vernon Bike Trail.

the river. The urban skyline of Rosslyn and other landmarks on the Virginia shore are visible to varying degrees depending on the season. These include the Pentagon, the Air Force Memorial, and the Iwo Jima Memorial. To the north, the view is dominated by Theodore Roosevelt Island, the Kennedy Center for the Performing Arts, and the adjacent Watergate Hotel. Perched on a hill in the distance is the Washington Cathedral.

Arlington Memorial Bridge is a prominent symbol of the city for Washingtonians, but it is likely less familiar to people outside the region, simply because it is not visible from the tourist areas. Although the argument could be made that photographs of it are readily recognizable as being located in Washington, D.C., this is because those photos almost always include the nearby Lincoln Memorial and Washington Monument in the frame. Take those away, and the bridge becomes much less familiar.

TIDAL BASIN BRIDGES

While hundreds of thousands of tourists stroll around the Tidal Basin to view the cherry blossoms each spring, it is likely that few of them know about the unusual engineering function performed by the Tidal Basin. It was not constructed just to make new parkland, golf courses, or spaces for memorials and cherry trees. Instead, the Tidal Basin and the bridges that connect it to the Potomac are an ingenious machine for keeping the Washington Channel open to shipping and boat traffic.

The Washington Channel is the narrow arm of the Potomac River between East Potomac Park and the Southwest Waterfront neighborhood. It is lined with marinas, houseboats, and yachts for Washington's well-to-do. If not engineered or maintained, the Washington Channel would quickly fill with sediment, making it inaccessible for pleasure boats and dramatically reducing the desirability of real estate along the Waterfront. To prevent this, the Tidal Basin operates as a gigantic water storage and release facility.

Between 1882 and 1897, the U.S. Army Corps of Engineers dredged the Potomac and then used the dredged sediment to fill in low-lying, swampy areas of what later became the Mall. This also created an artificial island that became East and West Potomac Parks, which in turn surrounded a body of water called Twining Lake. This artificial island separated the main channel of the Potomac River from the narrow Washington Channel.

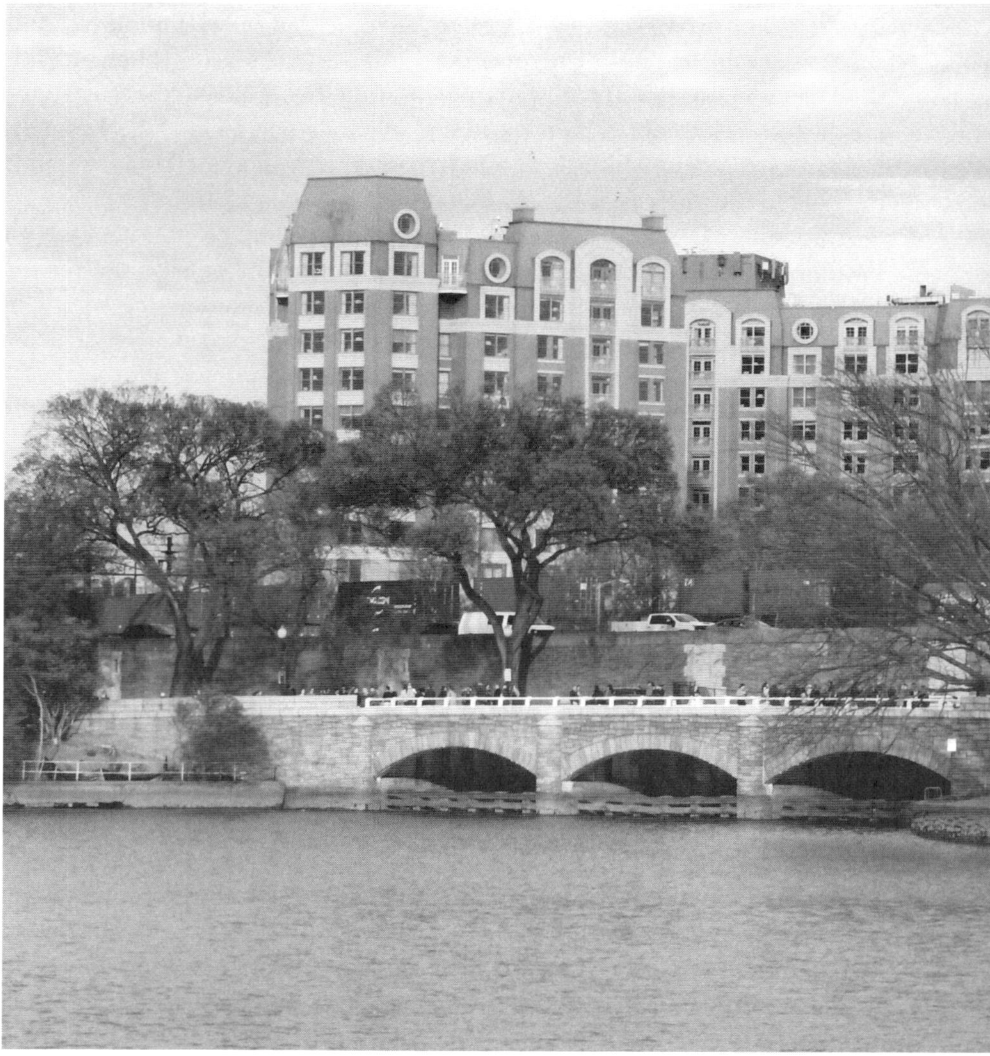

View of the Outlet Bridge at the Tidal Basin, during peak cherry blossom season.

To keep the Washington Channel free of sediment without the need to continually dredge, Peter Conover Hains, a military engineer, devised a plan to use the island as an engine to keep the Washington Channel clear of sediment by using the force of tides. The Outlet Bridge on the southern end of Twining Lake was the earliest of the three bridges now part of the Tidal Basin complex. It was constructed in 1889 and still operates. The bridge is

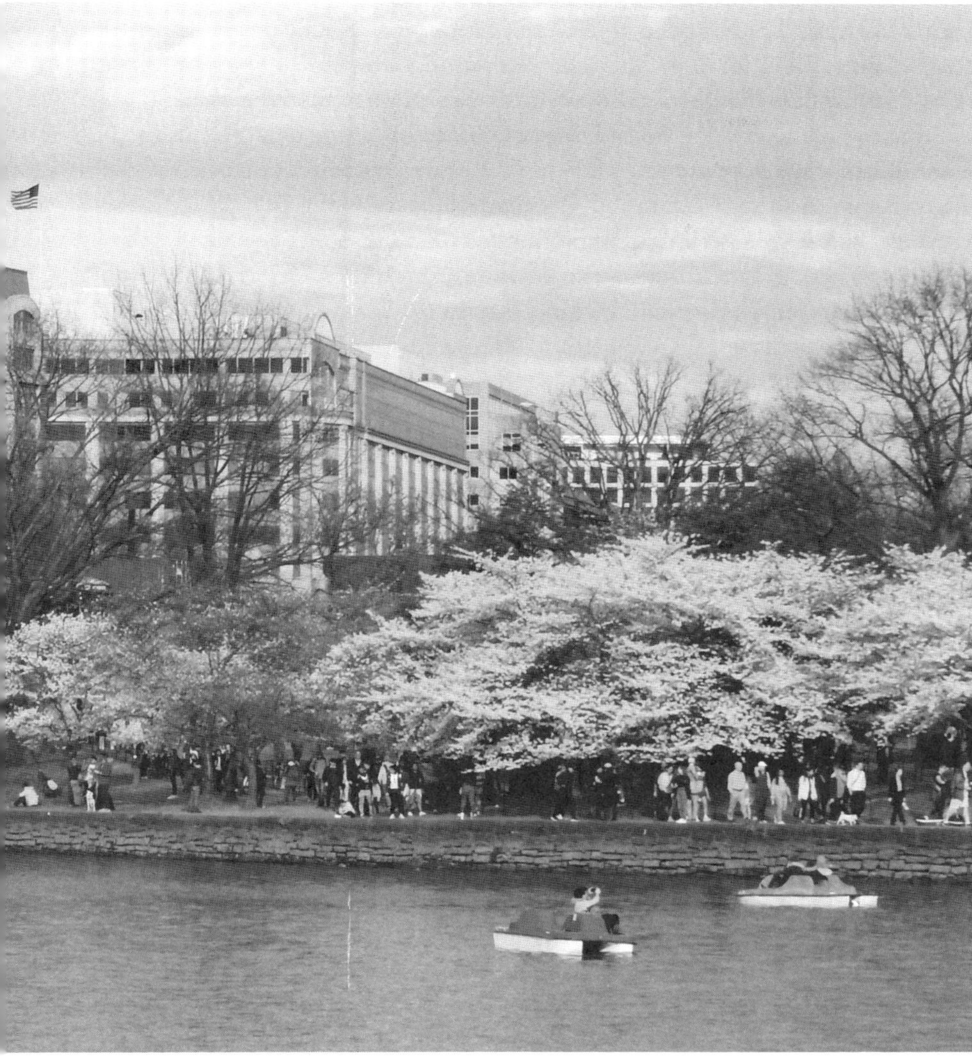

relatively inconspicuous. It is only about ninety feet long, sits about ten feet above the water, and is faced with reddish granite. It carries two lanes of traffic of Ohio Drive, the roadway that circles the Tidal Basin. It also has a pedestrian sidewalk that makes a complete circuit walk around the Tidal Basin possible.

The bridge structure is composed of arches, and the interior of each arch is fitted with wooden gates. The rising tide automatically closes the gates, trapping water in the Tidal Basin. Then, as the tide in the Washington

Channel drops, the gates in the Outlet Bridge open and the gates in the Inlet Bridge close, forcing the higher-elevation water in the Tidal Basin to flow out through the gates of the Outlet Bridge. Because the Washington Channel is relatively narrow, the large volume of water escaping from the Tidal Basin must flow quickly, allowing it to sweep sediment out of both the Tidal Basin and the Washington Channel. This cycle happens twice a day, keeping both free of sediment and open for pleasure boats.

At the time of construction, there were no parks, memorials, or other developments on the artificial island, so there was no need for a bridge over the outlet structure. By 1906, pedestrians had installed makeshift handrails and were unofficially using the outlet structure to cross the flats on foot. In the 1940s, the width of the bridge was expanded to carry traffic from Ohio Drive to Fifteenth Street and to formalize the pedestrian use of the bridge. Today, hundreds of thousands of tourists use the bridge to complete a circuit walk of the Tidal Basin during the National Cherry Blossom Festival. However, few of them know what is happening beneath their feet to maintain the Washington Channel for boat owners and Waterfront residents.

As late as 1894, engineers hoped that the Washington Channel could be kept clear using only the Outlet Bridge, without the need for controls on the inlet at the Potomac River on the western side of the Tidal Basin. However, this allowed the ebbing tide to flow out of both sides of the Tidal Basin, limiting the effectiveness of the system. Construction of an Inlet Bridge began in 1908. A similar system of tidal gates on the Inlet Bridge closes when the tide is dropping, forcing all the water in the Tidal Basin to exit through the Outlet Bridge. By channeling a larger amount of water through the Washington Channel, the water must move more quickly, and it is therefore more effective in removing sediment from both areas.

In addition to improving the function of the Tidal Basin, other things were happening in 1908 and 1909. At that time, automobiles began to use the roadways, increasing the need for bridges. The Inlet Bridge was intended, from the start, to carry cars and other traffic on Ohio Drive, completing the roadway circuit around the Tidal Basin. Another important factor in 1908 was that the McMillan Commission was in the process of developing the land surrounding the Tidal Basin into East and West Potomac Park, and the City Beautiful Movement was influencing the aesthetic design of bridges and other public structures throughout the city. As a result, the Inlet Bridge was designed to enhance the aesthetics of the developing park and memorials.

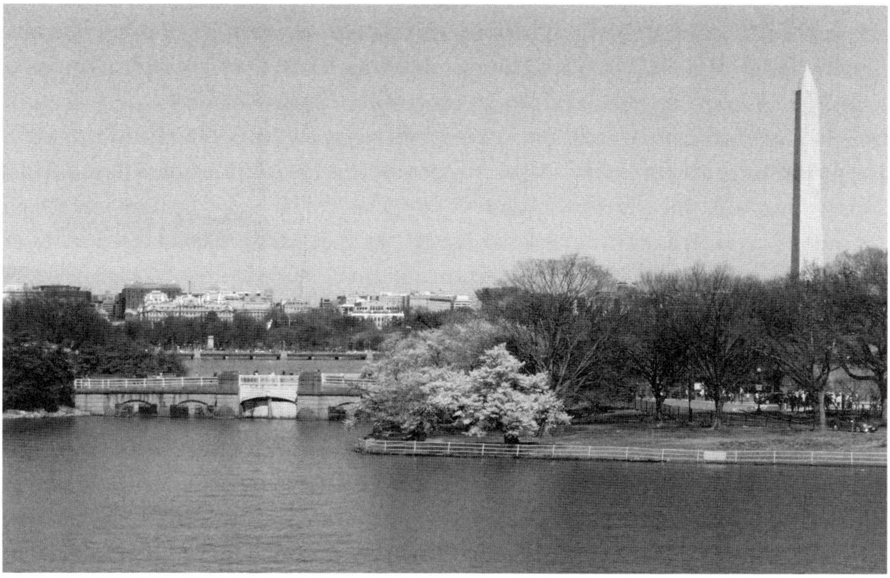

View of the Tidal Basin Inlet Bridge, Kutz Bridge, Washington Monument, and White House taken from sidewalk on the George Mason Bridge.

The bridge consists of three reinforced concrete arches. It is 184 feet long, with two traffic lanes and a sidewalk on each side. The central arch originally included a lock to allow passage of small boats between the river and the Tidal Basin, and the deck of the central arch could be raised to allow passage of larger boats. This mechanism, apparently never used, was removed and replaced with a fixed span in 1985.

The inlet crossed by the bridge connects the Tidal Basin to the Potomac River, making the bridge highly visible from both areas. The bridge was designed by Nathan C. Wyeth, the same architect who designed Key Bridge. The concrete is ornamental and has been scored to look like masonry. The balustrade on the two flanking arches is also concrete, with a geometric open-circle pattern, while the balustrade on the central arch is iron. The piers on either end of the central span have niches in them that contain bronze fountains, with the date of 1908 stamped into the concrete of one pier and 1909 into the other.

The circuit walk around the entire Tidal Basin, past all of its monuments, taking in its amazing views of the Washington Monument and, especially in spring, surrounded by cherry blossoms, is one of the most iconic experiences for tourists, visitors, and District residents. The circuit trail completed by the Inlet Bridge makes this possible. From the sidewalk on the bridge, the impressive Jefferson Memorial dominates the view to your right, but it is probably not even the main visual attraction. From the sidewalk, straight across the basin, the northern shoreline is lined with cherry trees, with the Washington Monument towering above them. This is one of the most photographable locations in the city.

Not to be overlooked is a third bridge at the Tidal Basin. This is Kutz Bridge, named for Charles Kutz, who was the commissioner of engineering for the District in the early 1900s. Structurally and decoratively, the Kutz Bridge is not particularly special. It is a low-profile steel-girder bridge sitting on concrete piers and seems a little out of place. The westbound lanes of Independence Avenue, on the southern side of the Mall, curve around an odd, oval-shaped protuberance on the northern end of the Tidal Basin. In 1943, the Kutz Bridge isolated this northern extension by cutting directly east–west across the northern end of the Tidal Basin. Once that was completed, the original alignment of Independence Avenue was made one-way westbound, and traffic on the new bridge became one-way eastbound on Independence Avenue. In this way, the bridge simplifies traffic patterns in and out of Southwest D.C.

The bridge has sidewalks on both sides, with the southern sidewalk offering a view of the Jefferson Memorial across the water, while the northern sidewalk gives a direct view of the Washington Monument. More importantly, the western end of the bridge is a prominent gathering place for tourists and visitors. The Martin Luther King Jr. Memorial is just a few feet from the end of the bridge, and the bridge is within a short walking distance of the Washington Monument, the World War II Memorial, the Korean War Veterans Memorial and the Lincoln Memorial.

Of special importance to many visitors, the western end of the Kutz Bridge was also the location of the first planting of cherry trees. One of the major attractions of the Tidal Basin, the trees were a gift to the United States from Tokyo in 1912. A bronze plaque, attached to a stone, commemorates the location.

FOURTEENTH STREET COMPLEX—LONG BRIDGE

The complex of bridges collectively referred to as the "Fourteenth Street Bridge" is one of the least attractive but one of the more historically important bridge locations in the District. With twelve traffic lanes, a Metro subway line, and the Virginia Railway Express, it is also the most heavily traveled bridge corridor bringing commuters into the city. The complex also provides a biking and pedestrian link between the Southwest Waterfront neighborhood and the Pentagon/Crystal City area in Virginia.

Because the complex involves five parallel bridges with different attractions, this discussion is split into two parts: one addressing the railroad bridge, and the second describing the adjacent highway bridges.

From its founding, the development of Washington, D.C., was influenced by the rivalry between the two port cities within its boundaries: Georgetown and Alexandria. Georgetown, situated near the free-flowing portion of the Potomac River and Rock Creek, offered water power to drive mills, as well as the promise of access to the Ohio River basin for the westward-expanding nation. With its deeper harbor and road connections to Virginia and the rest of the southern states, Alexandria was the better port.

The source of the rivalry was that Alexandria was closer to the Chesapeake Bay, and shipping to and from Georgetown required passing Alexandria on its way. This meant that anything Alexandria did to improve its own shipping industry—dredging, filling, clearing new channels, or putting obstructions such as bridges into the river—had the potential to interfere with Georgetown's shipping and business interests. Alexandria's position was not hurt by the fact that four of the first five presidents of the fledgling nation were Virginia natives, while Georgetown was part of Maryland. Throughout the District's early years, business interests in Georgetown often opposed Alexandria's proposals for river or harbor improvements, but the Alexandrians usually won the argument.

The key to the location and orientation of the Long Bridge, the earliest bridge in the complex that became the Fourteenth Street Bridge, can be found in the geographical history of the Federal City at the time of its founding. The port of Alexandria is situated southwest of the governmental core of the city, represented then, as now, by the Capitol. The L'Enfant Plan for the city included Maryland Avenue, which radiated out from the Capitol toward the southwest. Maryland Avenue reached the eastern shore of the Potomac River at Fourteenth Street, almost directly opposite Alexandria. Thus, the shortest traffic route between the Capitol and Alexandria was Maryland Avenue, needing only a bridge at that location to make the link complete between the new capital city and the Southern states.

The bridge was completed in 1809, becoming the second crossing of the Potomac following the bridge at Little Falls. According to some sources, the origin of the name "Long Bridge" came from the contrast between the relatively short bridge at Little Falls and the mile-long bridge—the longest in the United States at the time—connecting Maryland Avenue to Alexandria.

The original bridge at the location was made of wood and sat only ten feet above river level. It was a toll bridge, privately owned by the Washington Bridge Company. The bridge had drawspans on each end to allow ships to reach Georgetown. However, the drawspans were narrow, and large vessels under sail had difficulty navigating through to reach Georgetown.

On August 24, 1814, President James Madison fled the city in advance of the burning of the Capitol and White House by the British. He escaped to the northwest using the Georgetown Ferry, while other government and military officials escaped over the Long Bridge. The Americans then burned the western end of the bridge to stop the British from following them into Virginia. The next day, the British burned the eastern end of the bridge to stop the Americans from reentering the city and pursuing them as they returned to their ships on the Patuxent River. The bridge was reconstructed and reopened by 1818.

Much of the bridge's history during the nineteenth century was an ongoing battle against flooding and ice floes. The first big flood was in 1831, resulting in the federal government taking over after the bridge's private owner filed for bankruptcy. The bridge was reconstructed again as a low-profile wooden bridge with drawspans, which reopened in 1835. The opening was a major event, with President Andrew Jackson and his cabinet walking across, but then choosing to take carriages back across the mile-long bridge. The wooden bridge was then destroyed or damaged by floods or ice in the years 1841, 1856, 1860, 1863, 1866, 1867, 1870, 1881, 1887, and 1889.

Having been constructed before the invention of railroads, the Long Bridge originally carried only foot, horse, and carriage traffic. However, even after railroads reached the District in the 1830s, political infighting still kept the bridge from being used for railroad traffic. From its construction until 1847, the bridge was entirely within the boundaries of the District, but the retrocession of 1847 resulted in the southern end being absorbed back into Virginia.

Several railroad companies vied to use the structure to connect Alexandria to Washington by rail. In the 1850s, the B&O Railroad attempted to connect the bridge to its New Jersey Avenue Station near the Capitol. Tracks were even laid to the end of the bridge. However, the Virginia legislature opposed using rail on the bridge, so goods and people were shipped by rail to the end of the bridge, loaded onto carriages and wagons, and carried across the Potomac. The goods were then reloaded onto another railroad to continue the journey to Alexandria.

During the Civil War, the bridge served as the connection point between Washington and the Virginia battlefields. In April 1861, when Virginia seceded from the Union, federal troops immediately occupied the bridge and the port of Alexandria and quickly laid tracks across the bridge to support the Civil War effort. However, the bridge was found to be too weak to support rail traffic, and the railcars had to be lightly loaded and pulled by horses. A second bridge was completed by a private company in 1863, but shortly afterward, the original bridge was damaged by the weight of a train. The federal government purchased the second bridge and laid rails on it, leading to the new name of Railroad Bridge. The original bridge, repaired and used only by horses and carriages, became designated as the Turnpike Bridge. Meanwhile, the second bridge added to the obstruction of water flow in the river, resulting in increased siltation, which caused additional problems for the businesses in Georgetown.

Part of the eventful history of the Long Bridge derives from its role in crossing the jurisdictional boundary between the District and Virginia. A famous poem by Sara Jane Lippincott, titled "The Leap from Long Bridge," relates the story of a nineteen-year-old slave girl who, in 1848,

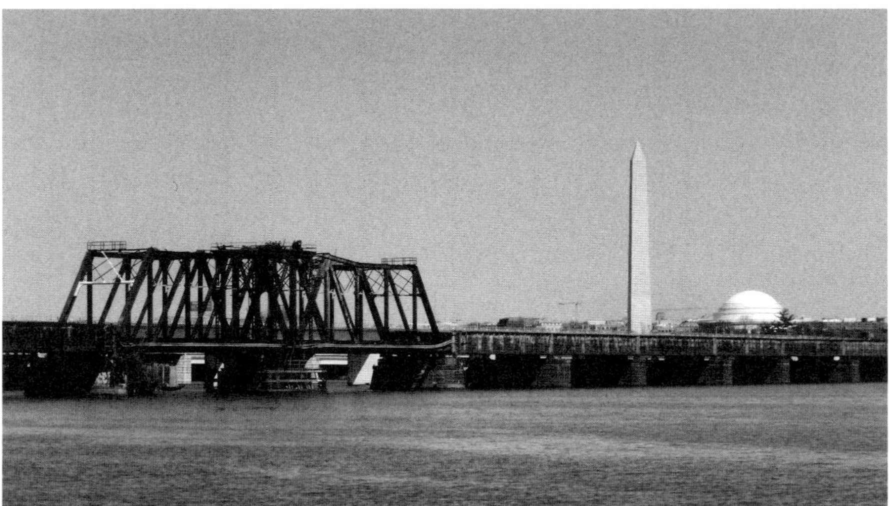

The 1904 Long Bridge, showing the rotatable span that allowed shipping traffic to support industry in Georgetown.

was to be transported to Georgia for sale to a cotton plantation. She escapes, attempting to reach the woods on the Virginia side of the river, but is caught on Long Bridge. Rather than be captured, she jumps to her death in the river.

There are also stories of the period from 1890 to 1900, when the District prohibited gambling. This caused certain unsavory elements of society to convert the western end of the bridge into "Jackson City," a community of saloons, pool halls, and opium dens. The bridge, with its inadequate lighting and hiding places among the wooden trusses, became a refuge for criminals.

The massive flood in 1889 and the need to revitalize the area occupied by Jackson City led to the plan to replace the Railroad and Turnpike Bridges in the early 1900s. The new Railroad Bridge opened in 1904, and is still owned and used by CSX Transportation, as well as Amtrak passenger trains and Virginia Railway Express commuter trains. After more than a century of being called the "Railroad Bridge," the name "Long Bridge" came back into fashion in the 1980s and is the name most commonly used by residents.

Like Roosevelt Bridge, the Long Bridge is a relatively unattractive industrial-looking steel-girder bridge, but it is at least partly hidden from view of most of the monuments by the later highway bridges that make up the Fourteenth Street Bridge complex. It does, however, pass directly over scenic Ohio Drive and crosses East Potomac Park on its eastern end and the Mount Vernon Trail on its western end. There is no pedestrian or bike access to the bridge or historic information placards nearby describing its importance in the District's history. However, there is plenty of information about its eventful history online, and history buffs, whether local or visiting, can use the hiking and biking trails on either end to visit and view the structure.

FOURTEENTH STREET COMPLEX—HIGHWAY BRIDGES

The Fourteenth Street Bridge's complicated origins are found in the history of the Long Bridge discussed above. By the end of the Civil War, the historic Long Bridge had morphed into two bridges: the Railroad Bridge and the parallel Turnpike Bridge. By the late 1890s, both bridges required replacement, with the new Railroad Bridge opening first in 1904. The new Highway Bridge opened in 1906 and was a steel through-truss bridge, sixty-one feet wide, with two sidewalks and electric lighting.

The 1906 Highway Bridge quickly became too small and weak for the amount of truck and bus traffic, and reinforcement was needed as early as 1928. In 1946, the District Highway Department proposed two new bridges, each four lanes wide. The first of these bridges opened in 1950 and was named the Rochambeau Memorial Bridge after the French Revolutionary War hero. It was a plain utilitarian highway bridge with a bascule drawspan in the center and a stone-faced operator's tower near the western end. The new bridge was used entirely for inbound traffic, while the 1906 bridge continued to be used for outbound traffic.

The second bridge was completed in 1962 and was also a plain utilitarian bridge, although it has a narrow sidewalk that can be used by pedestrians and cyclists to travel across the Potomac from Virginia to the Tidal Basin. This bridge was named for George Mason, a Virginian who was one of the nation's founders. The original plan for a drawspan on the second bridge was eliminated to lower the cost. This had substantial consequences for other bridges on the river, because it closed off shipping access farther inland. This effectively eliminated the need for the moveable spans on Arlington Memorial, Rochambeau, and Long Bridge.

Following completion of the George Mason Bridge, the 1906 Highway Bridge was removed in 1967. The trusses were floated downriver through the final opening of the drawspan of the Rochambeau Bridge to Dahlgren, Virginia. There, the bridge trusses served as bomber target practice for the military.

The third traffic bridge in this complex opened in 1972 and was constructed directly adjacent to the Mason Bridge. Since 1977, all three highway bridges have been designated as Interstate 395, and they also carry US Route 1. When it was opened, the third highway bridge was not given a name.

One of the most memorable events for Washingtonians in the early 1980s was the snowstorm of January 13, 1982, when an Air Florida jet taking off from National Airport struck the Rochambeau Bridge, killing seventy-four passengers and four people on the bridge. One of the passengers, Arland D. Williams, helped rescue several people from the water before he himself drowned. Following this tragedy, the 1950 Rochambeau Bridge was renamed for Williams, and the Rochambeau name was transferred to the unnamed 1972 bridge.

The final bridge added to the complex was the Charles Fenwick Bridge, which carries Metro Yellow Line subway trains. The Fenwick Bridge, situated between Long Bridge and Arland Williams Bridge, opened in 1983.

Today, the five bridges making up the complex have limited visual appeal. They are primarily utilitarian, midcentury steel-girder bridges. They are all low-profile, sitting on piers with no substantial superstructure, so they are not visible from a long distance, and there are no towers or cables that can act as works of modern public art. Only one of them, Long Bridge, has an interesting historic background.

FRANCIS CASE BRIDGE

Most of the circumference of the Tidal Basin is part of West Potomac Park, which forms the Tidal Basin's eastern, northern, and western shores. The southern shore of the Tidal Basin, where the Thomas Jefferson Memorial is located, is part of East Potomac Park. East Potomac Park is an island, separated from West Potomac Park by the narrow channels crossed by the Inlet and Outlet Bridges. The five bridges of the Fourteenth Street Bridge complex traverse from Virginia over the Potomac and onto the island of East Potomac Park.

Constructed in 1962, the Francis Case Memorial Bridge acts as an extension of the Fourteenth Street complex by continuing eastward across the Washington Channel into the Southwest Waterfront. In doing so, the bridge completes a pedestrian and bicycle corridor between the District and Virginia. Of the five bridges making up the Fourteenth Street complex, only the George Mason Bridge has a sidewalk, which is on its northern side. The sidewalk ends in East Potomac Park, with a ramp carrying pedestrians and cyclists down to the George Mason Memorial, Jefferson Memorial, and other attractions in East Potomac Park. Similarly, the Case Bridge carries a sidewalk over the Washington Channel from Benjamin Banneker Park in the Southwest Waterfront to a ramp leading down to the tennis center in East Potomac Park. While there is no continuous trail in East Potomac Park connecting the sidewalk from Mason Bridge to the sidewalk on Case Bridge, the link can be made by navigating sidewalks within the park. The connection is convoluted, but it does exist.

By day, the Case Bridge is a relatively unattractive, midcentury steel-beam bridge. The piers, clad in decorative stonework, are reasonably attractive from viewing points within East Potomac Park and the Southwest Waterfront. However, at night, the bridge comes alive with colored lighting, installed in 2016. The edge of the deck facing southeast toward the Waterfront is traced with a solid, dark-blue line of lights, but the more appealing feature is that each of the beige-colored, stone-faced piers is washed in colors ranging from dark blue or purple to a light lavender. The textured surface of the stonework on the piers is rough, so the high angle of the lighting fixtures creates an attractive mixture of deep shadows and bright highlights. All of this is reflected in the still waters of the Washington Channel and presented against a backdrop of the Washington Monument and modern condo, restaurant, and shopping developments at the Waterfront. For pedestrians, the best place to get an unobstructed view of the lighting is from Ohio Drive in East Potomac Park.

WOODROW WILSON BRIDGE

The Woodrow Wilson Bridge, which carries the Capital Beltway across the Potomac River, barely qualifies as a District bridge. As the crow flies, the bridge sits more than six miles south of the Capitol Building, and that crow would be flying most of this distance over open water. By highway, the driving distance from the Capitol to the bridge is more than nine miles. The bridge is well-known to local residents as a connection between Maryland and Virginia, with a "Welcome to Maryland" sign on one end and a "Welcome to Virginia" sign on the other. Also, the bridge is almost completely invisible from within the District. It is barely visible from Hains Point, the southern tip of East Potomac Park. From a further distance, it is visible from the Washington Monument's elevated viewing platform.

Despite the bridge's distal location, a few people know that, for a short distance near the Virginia shoreline, it enters and immediately leaves the District's boundaries. There are no signs informing drivers or pedestrians. The crossing occurs over water, and there are no exits off the bridge into the District. The total length of the bridge within the District boundaries is only about four hundred feet, a distance that cars traveling sixty miles per hour will cover in about five seconds.

This brief and unannounced excursion into the District is not enough to justify inclusion of the bridge in a book on the attractions of the city's bridges. However, several other features of the bridge not only qualify it for inclusion but also make it one of the most historically important bridge locations in the entire city.

As discussed in the introduction, the approach to bridge design changed in the mid-twentieth century, as community uses and aesthetics were eschewed in favor of utility and speed. Most of the District's highway bridges over the Potomac and Anacostia were constructed during this time—and it shows. Roosevelt Bridge and the Fourteenth Street Bridge complex remain as a reminder of how unappealing midcentury highway bridges could be. Until the early twenty-first century, the Roosevelt and Fourteenth Street Bridges were joined by several similarly unattractive midcentury highway bridges, including the original Frederick Douglass and Eleventh Street Bridges over the Anacostia and the original Woodrow Wilson Bridge over the Potomac.

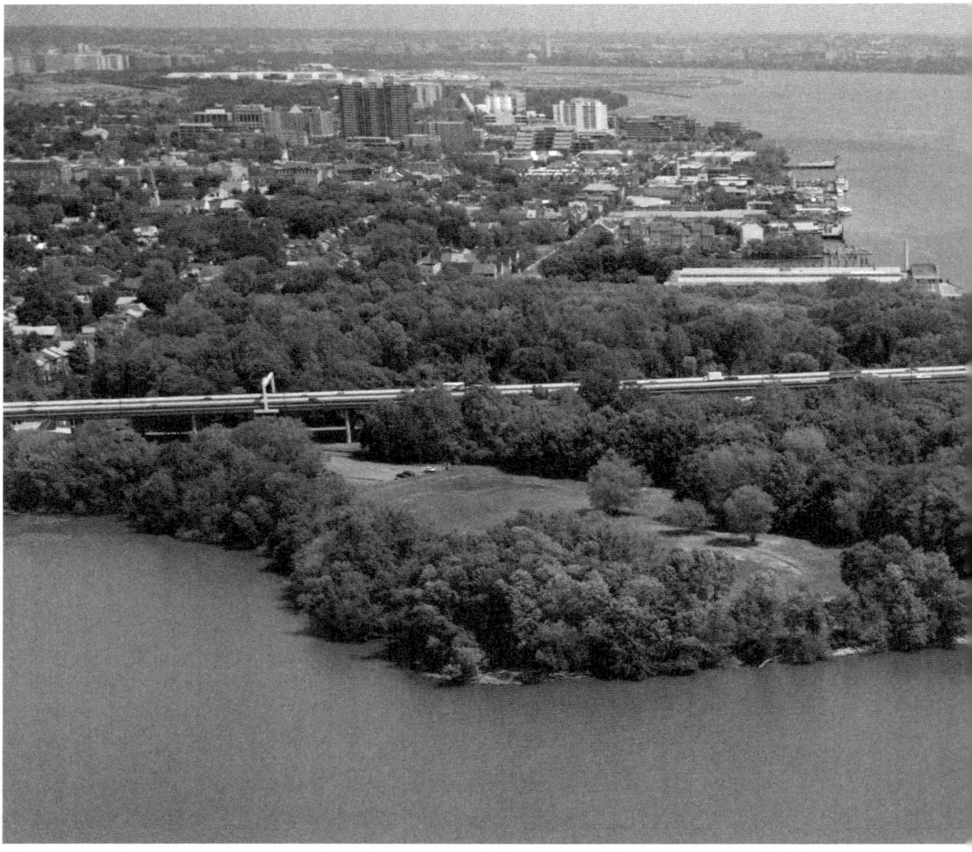

The introduction also mentions how the approach to bridge design changed in the late twentieth century to incorporate aesthetics and community-friendly features. This was done with construction of sculptural cable-stayed and pedestrian bridges and preservation of obsolete historic bridges as parks and trails in hundreds of cities all over the United States.

One city came late to this party: Washington, D.C. This is one reason why, despite many cities and rivers having had books written about their bridges, the District has not been one of them—until now. But this late arrival can be excused because the District has done a stellar job over the past twenty years of catching up to other cities. And the first location to bring the District,

Aerial view of the original Woodrow Wilson Bridge and Jones Point Lighthouse. *Library of Congress.*

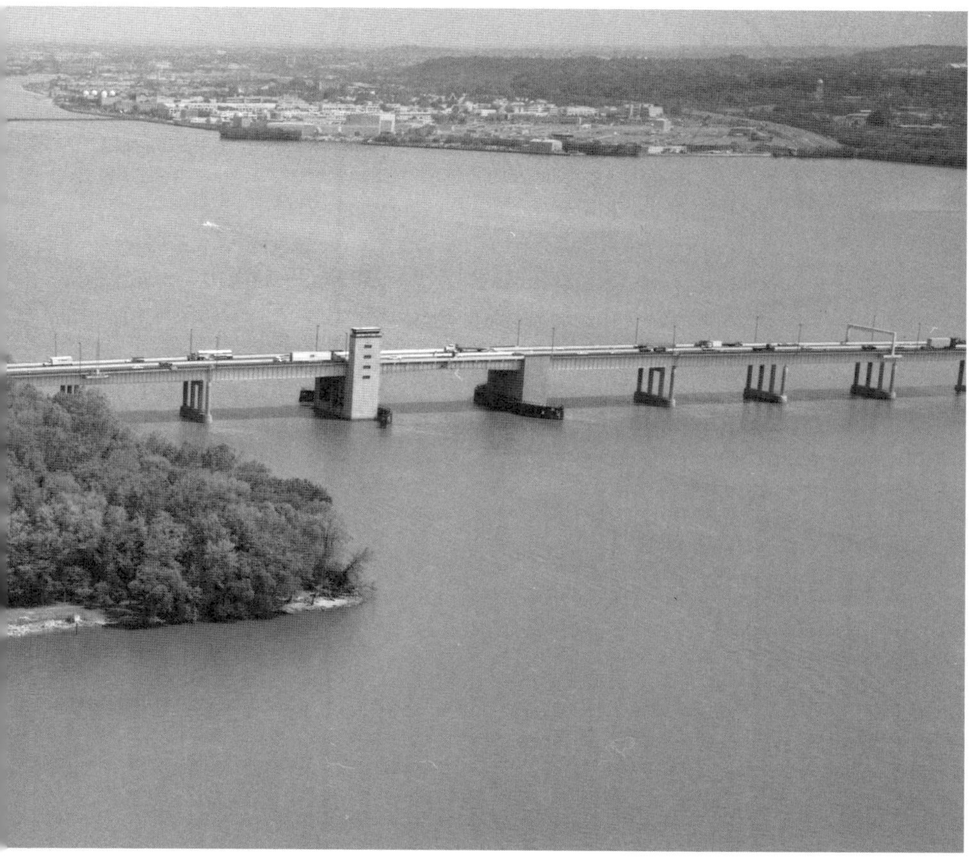

Maryland, and Virginia into twenty-first-century bridge design was the Woodrow Wilson Bridge.

The federal and local governments did not spend billions of dollars to demolish the original Wilson Bridge and construct the new one in order to replace an unattractive bridge with a work of modern art. This is why these infrastructure changes in our cities happen so slowly. It is often difficult to justify spending tax dollars on aesthetics. However, once infrastructure needs replacing because of obsolescence, it is relatively inexpensive to incorporate aesthetics and community functions into the new design.

The original Woodrow Wilson Bridge opened in 1961. At that time, the Capital Beltway was a six-lane highway, so the jurisdictions constructed a six-lane bridge. This was a big mistake, because traffic volumes boomed and the Beltway had expanded to eight lanes by the 1970s. This led to major problems as eight lanes of traffic were squeezed onto the six-lane bridge. These traffic jams were made much worse because the low-profile bridge also had a drawspan, which opened frequently to allow ships carrying materials into Alexandria.

By the early 2000s, the bridge needed to be replaced. The replacement would be the first major bridge constructed in, or near, the District in almost forty years. It would also be the first bridge constructed in this new era of bridge design in which aesthetics, recreation, and other community-friendly functions were important considerations. And as it turned out, it was just the beginning of this new era in the District, with the Eleventh Street and Frederick Douglass Bridges constructed shortly afterward and more still to come.

The first design challenge for a replacement bridge was its height above the water. One of the major problems to be solved with a new bridge was the elimination of traffic jams caused by the opening of the drawspan. Constructing another low-profile bridge with a drawspan was not an option. This suggested that a high-profile bridge, such as a traditional suspension bridge or a modern cable-stayed bridge, would be needed. However, local residents, especially those in historic downtown Alexandria just a mile away, opposed the idea because it would intrude on scenic river views. The compromise was to construct a bridge with a drawspan twenty feet higher than the original bridge so that the frequency of drawspan openings could be reduced.

In retrospect, this compromise was ironic for two reasons. At the time of its design and construction in the early 2000s, there was only one remaining industry inland from the bridge that regularly received materials via shipping.

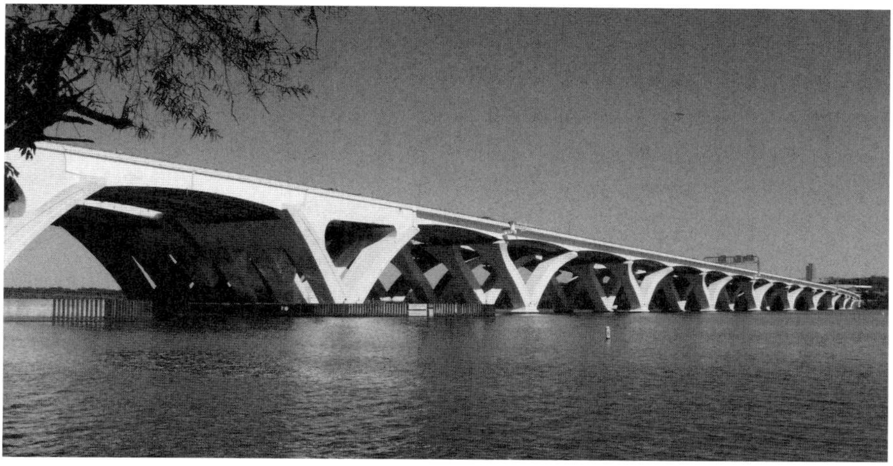

The 2007 Woodrow Wilson Bridge, the first sculptural bridge constructed near Washington, D.C.

This was the Robinson Terminal in Alexandria, which received frequent shipments of newsprint for *The Washington Post*. And everyone knows what has happened to the newspaper industry. The new bridge with its drawspan opened in 2007, and the Robinson Terminal ceased operations in 2013.

The second irony was that many residents and businesses in Alexandria objected to the visual intrusion of a high-profile bridge. Given the appearance of most other large-scale bridges in the District at that time, fear of another unappealing bridge constructed within sight of the historic downtown was justifiable. However, the aesthetics of bridge design had changed. The piers of the bridge were designed with a sleek, modern, curved *V* shape, resulting in a gorgeous bridge that the community can be proud of. The bridge has won numerous awards for the innovation and beauty of its design, and it is a benefit to the community that it sits high enough to be prominently visible, if not from downtown Washington, at least from Alexandria.

Another example of foresight was the incorporation of recreational features into the bridge. At the time of construction, the area on the eastern end of the bridge was largely undeveloped. In 2007, incorporating a wide sidewalk, isolated from the traffic lanes, crossing two miles of bridge to an area lacking any attractions, seemed overly optimistic. Today, National Harbor, located directly off the eastern end of the bridge in Maryland, is an enormous complex of hotels, shops, and restaurants with a casino, a convention center, and other attractions—all accessible to Alexandria residents on foot or bicycle.

In addition, the fifteen-foot-wide sidewalk is much wider than needed if it were just used by pedestrians to cross from one side to the other. Instead, it was deliberately designed to be wide enough to support recreation, including cyclists, roller-bladers, joggers, and hikers. There are even wider, parklike gathering places along the sidewalk, in the middle of the Potomac River, that include benches, historic information placards, and observation telescopes with views of the downtown monuments. The sidewalk connects to the seventeen-mile-long Mount Vernon Trail, which passes through Jones Point Park directly underneath the western end of the bridge. On the eastern end, the sidewalk connects to an overpass crossing the Capital Beltway to the National Harbor development on the bridge's southern side. The overpass is designated as part of Potomac River Waterfront Park. It is pedestrian-only and extensively landscaped as another recreational space situated directly above the traffic lanes of the Beltway.

Probably known to very few Washingtonians, Jones Point Park is an unexpected treat. Because the bridge is elevated over the river, the highway, supported by the V-shaped piers, extends about half a mile into Virginia. This left a "down-under" space beneath the highway ramp that was developed into a lovely recreational park. Beneath the highway ramp, the Mount Vernon Trail passes by basketball courts, restrooms, historical displays and information placards, open fields, and fishing piers.

However, Jones Point Park is also an important attraction for geography and surveying geeks. It is here that the first stake was driven and the first stone set that established the boundaries of the Federal City. George Washington selected the city's location on the Potomac and directed that it be established as a 100-square-mile, diamond-shaped property, ten miles on each side and oriented exactly forty-five degrees away from a north–south axis. In 1791, Andrew Ellicott and Benjamin Banneker were hired to survey and mark the boundaries. This required picking a single spot as the starting point and establishing its latitude and longitude precisely. Once this was done, the exact location of the remaining boundary lines was established virtually by the forty-five-degree, ten-mile-per-side decision. The lines just needed to be marked on the ground. Jones Point, south of Alexandria, was chosen as the starting point.

Beginning in February 1791, Banneker spent two months establishing the precise latitude and longitude of the starting point at Jones Point using astronomical measurements and detailed calculations. Banneker and Ellicott then completed a rough survey of the forty-mile-long boundary line. On April 15, 1791, the Jones Point location was marked by the setting of a

Boundary Stone in a ceremony by the Alexandria Masonic Lodge. This original stone was replaced in 1794 and is thought to have been buried beneath a fence at the nearby St. Mary's Church graveyard.

Ellicott and his team, now working without Banneker, proceeded to complete a formal survey of the boundaries. They marked the boundaries by cutting a twenty-foot-wide swath and setting another stone every mile for a total of forty Boundary Stones. The sides of the stones were inscribed with the words "Virginia," "Maryland," or "Jurisdiction of the United States," along with the mile number. The fourteen stones in Virginia were set in 1791, and the twenty-six stones in Maryland were set in 1792. As of 2016, thirty-six of those stones were still in place and can be visited by history and geography buffs. Three others have been replaced, and one location is marked only by a plaque.

One Boundary Stone, though, is more important than the others: the stone at Jones Point, the first location established on the boundary line. Placed in 1794, it can be found just off the sidewalk on the back side of the Jones Point Lighthouse, a short walk south of the western end of Woodrow Wilson Bridge. Unfortunately, the stone, set beneath a glass cover, is barely visible or legible. However, other stones with visible engravings mark the original District-Virginia boundary and the District-Maryland Boundary in the park. Interestingly, the District-Virginia stones no longer mark a District-Virginia boundary because that part of the District, originally part of Virginia, was returned to Virginia as part of the retrocession of 1847.

4

C&O CANAL

In 1785, George Washington helped form the Potomac Company to improve the river for navigation. The early efforts were limited to constructing short canals and locks to bypass the rapids and falls near Georgetown and Harper's Ferry. Ultimately, the more ambitious goal was to connect the interior of the continent, which was rich in natural resources, with the well-populated areas of the Atlantic coast and, through the Atlantic coast harbors, with the foreign markets of England and Europe. Following the valley of the Potomac west, only a short connection across the Eastern Continental Divide would have been needed to allow shipping to access the Casselman River, which flowed north into the Youghiogheny and Monongahela Rivers and then, at what later became the city of Pittsburgh, the Ohio River. From the Ohio River, there was shipping access to the entire interior of the continent, including the Missouri River, the Mississippi River, and all the way to New Orleans and the Gulf of Mexico.

In 1828, construction of the C&O Canal began along a route on the northeastern side of the river, in Maryland. However, the canal was never completed west of Cumberland. The intensive period of canal construction in the early nineteenth century was known as "Canal Mania," but it was quickly followed by the development of railroads in the 1830s. Although the construction of the B&O Railroad quickly rendered the canal obsolete, 184 miles from Cumberland to the confluence of Rock Creek and the Potomac at Georgetown were completed and operated until 1924.

The C&O Canal is well-known to Maryland and District residents today because large parts of it are still intact and preserved as a historical park. After decades of neglect and disintegration, sections in and near the District were restored and rewatered and are now a major tourist attraction and recreational destination as the C&O Canal National Historical Park. The entire length of the towpath is still intact, stretching 184 miles from Cumberland to Georgetown. In addition to serving as a historical reminder, the towpath is one of the more popular long-distance hiking and biking trails on the East Coast.

CABIN JOHN (UNION ARCH) BRIDGE

The Cabin John Bridge, also known as Union Arch, is actually in Maryland and not the District, and is also not directly associated with the C&O Canal. However, it is located within easy walking distance from the canal and only a few miles from the District boundary, and was of enormous importance in the development of the city. Therefore, it is an important bridge to include in a tour of historic D.C. bridges.

The bridge crosses Cabin John Creek, a tributary to the Potomac. Despite it commonly being referred to as an aqueduct, it is not part of the system of water-filled troughs that carried canalboats over tributaries. Instead, it carries a massive water supply pipe from farther upstream on the Potomac River into the city.

The bridge was constructed as part of the effort to provide a stable water supply for the growing Federal City. A fire that destroyed the Library of Congress in 1851 highlighted the need for a reliable high-volume water supply, so planning began for the 18.6-mile-long Washington Aqueduct. Construction began in 1853 but was interrupted by the beginning of the Civil War. Construction resumed, and the structure was completed in 1863. The bridge does not carry a normal pipeline but encloses a nine-foot-diameter, brick-lined conduit for the transport of the water. The aqueduct still operates more than 150 years later, supplying the city with water from the Potomac at Great Falls.

The bridge's most remarkable feature, other than its history, is its enormous length and height. At the time of construction, it was the longest single-span stone bridge in the world at 220 feet long. The center of the arch is eighty feet high above Cabin John Creek and the adjacent Cabin John Parkway.

The bridge's size made it one of the most prominent tourist sites in the region in the late nineteenth century. Rosa Bobinger, the wife of one of

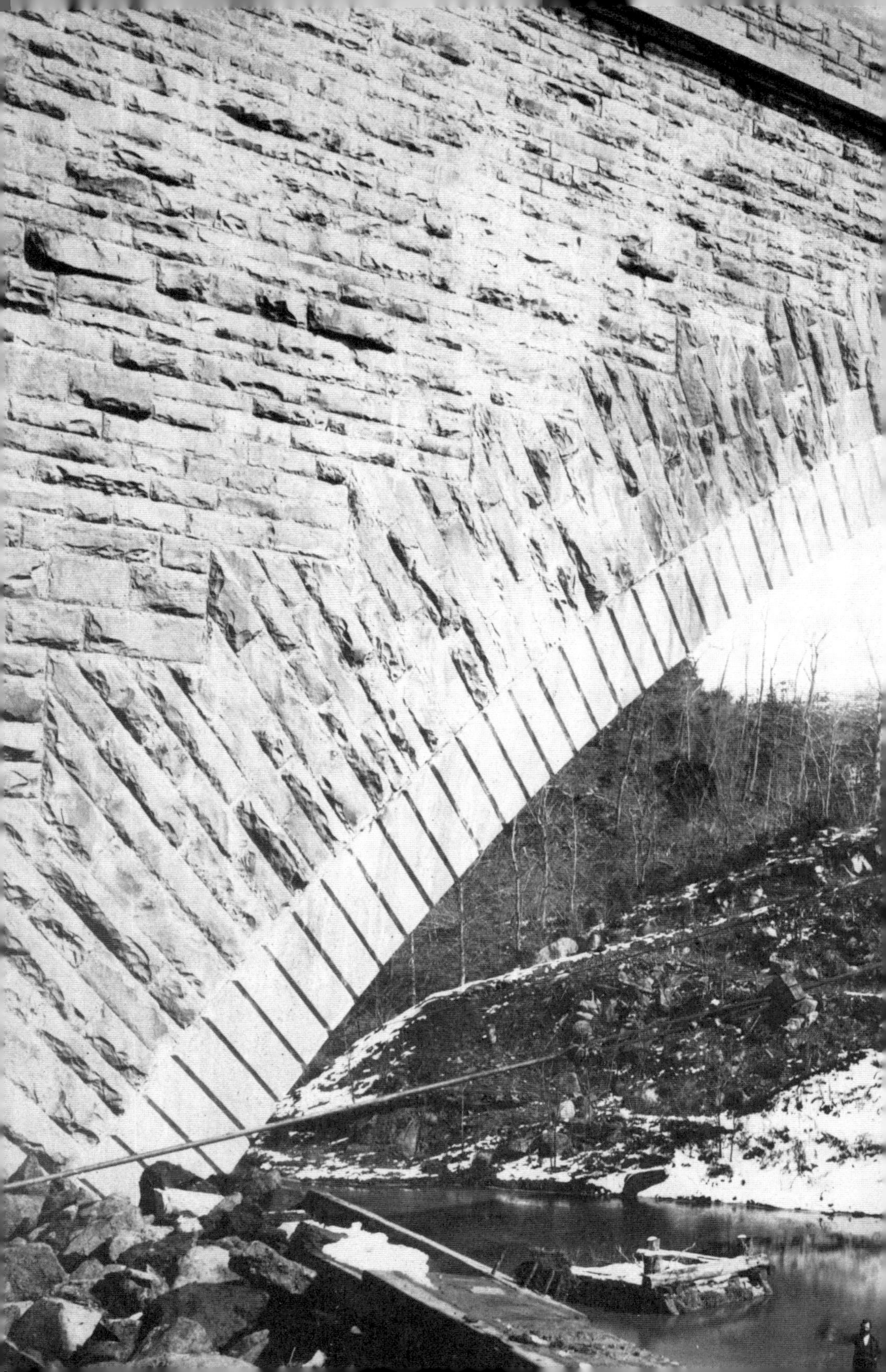

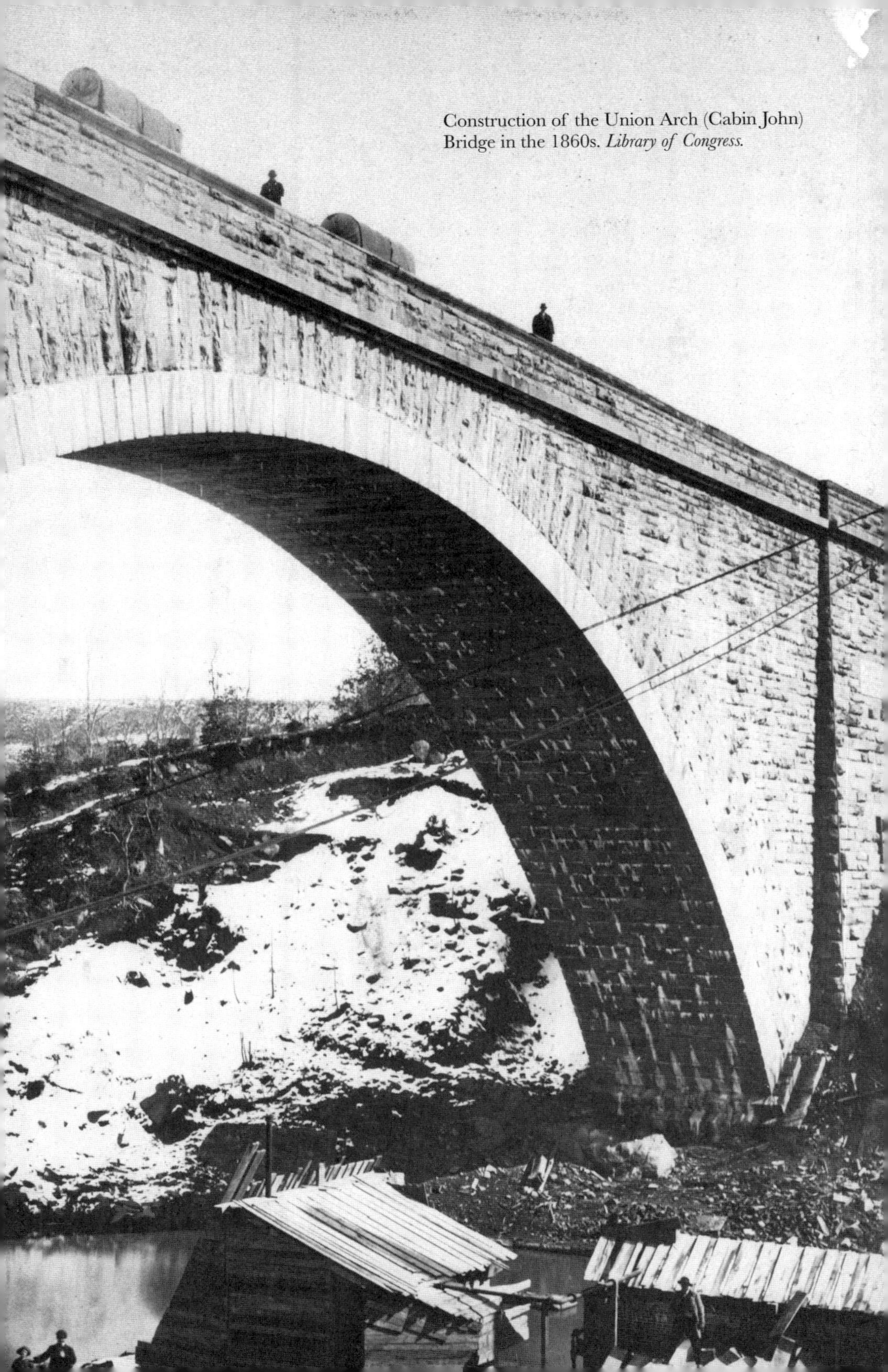

Construction of the Union Arch (Cabin John) Bridge in the 1860s. *Library of Congress.*

the stonecutters, set up a business selling meals to the construction workers. Realizing the bridge was a tourist attraction, she and her husband, Joseph, purchased the one-hundred-acre parcel of land at the northern end of the bridge in 1870. Then, as now, a scenic road followed the top of the Potomac Palisades from Georgetown out to the Maryland suburbs. Known today as MacArthur Boulevard, it was originally called Conduit Road, after the aqueduct that delivered water to the city. Conduit Road was a favorite route for an afternoon excursion away from the city to see the engineering marvel of the bridge and perhaps enjoy a meal and spend the night at Rosa and Joseph's Cabin John Bridge Hotel.

Under the management of Rosa Bobinger's sons, the hotel was expanded into an opulent, Gilded Age pleasure palace and gardens by the turn of the century. The District's well-to-do residents would come out to see the bridge and enjoy the scenery and cooler weather offered by the higher elevation and surrounding forests.

By 1896, the Foundry Branch Trolley Line, parallel to Conduit Road, had connected Georgetown to an amusement park at Glen Echo and then continued north to the southern end of the famous bridge. The Bobinger sons constructed a narrow cast-iron footbridge connecting the trolley station across Cabin John Creek to their hotel, where business boomed until the 1920s, when Prohibition caused a decline. The hotel closed in 1926 and then burned in 1931, still packed with antiques, china, and silver from its better days.

Even for local residents who know nothing of this history, the bridge remains a famous landmark due to its massive size. The area is hilly and wooded, so even though the bridge is large, it is not visible from a great distance. However, it is well-known to tens of thousands of local commuters who pass beneath it every day, because Cabin John Parkway is the link between Canal Road in Georgetown and the Capital Beltway leading into the Maryland suburbs.

From the sides or the top, there is no indication that a massive, brick-lined conduit is enclosed within the structure. Instead, it looks like a normal bridge, albeit an immense one. It even carries a roadway like a normal bridge, consisting of a single lane of MacArthur Boulevard. This makes the bridge a hindrance to traffic at times, because MacArthur Boulevard is a two-lane road, and the two-way traffic is controlled by traffic lights at each end of the bridge. The narrow width of the bridge, one traffic lane plus a single narrow sidewalk, seems discordant given the enormous height.

The bridge can be accessed on foot or bicycle by the asphalt trail that parallels MacArthur Boulevard and crosses the bridge. Cabin John Local Park is just off the northern end of the bridge and has ample parking. Other than driving beneath it at fifty miles per hour, the best way to view the bridge's profile is to follow the rustic hiking trails from the parking lot down into the valley along the creek.

ARIZONA AVENUE TRESTLE

Throughout the late 1800s, the B&O Railroad was thwarted in its attempts to either gain access to the Long Bridge or build its own bridge over the Potomac, in order to allow its rail system to expand into the Southern states. In 1892, plans were developed for construction of a spur off the main B&O Railroad line between Baltimore and Washington, which would have crossed a new bridge at Little Falls. For economic reasons, this never occurred. However, the railroad did complete construction of the Georgetown Branch Railroad—a spur off the main line—connecting Bethesda to Georgetown, in 1909.

Completion of the Georgetown Branch Railroad required a new bridge to carry the rail line across the C&O Canal and adjacent Canal Road. This was accomplished by disassembling two existing railroad bridges in Pennsylvania, transporting them to the C&O Canal, and reassembling them in their current location at the intersection of Arizona Avenue and Canal Road a few miles northwest of Georgetown. Historicbridges.org attributes the bridges to the Keystone Bridge Company of Pittsburgh, Pennsylvania, and their original construction date to the mid-1880s, based on the truss type and a "Carnegie" stamp on the girders. The website reports that the bridges were originally constructed in Ridley Park, Pennsylvania, before they were disassembled and moved to Washington.

Of the two separate trusses, one crosses only the C&O Canal, and the other crosses only Canal Road. The two trusses meet at a pier situated in the narrow median between the canal itself and Canal Road, forming a single bridge crossing composed of two trusses that are not only different structural types but also very different sizes. The C&O Canal segment is a Whipple truss, while the Canal Road segment is an unusual skewed Pratt truss structure that is only about half as high as the other truss.

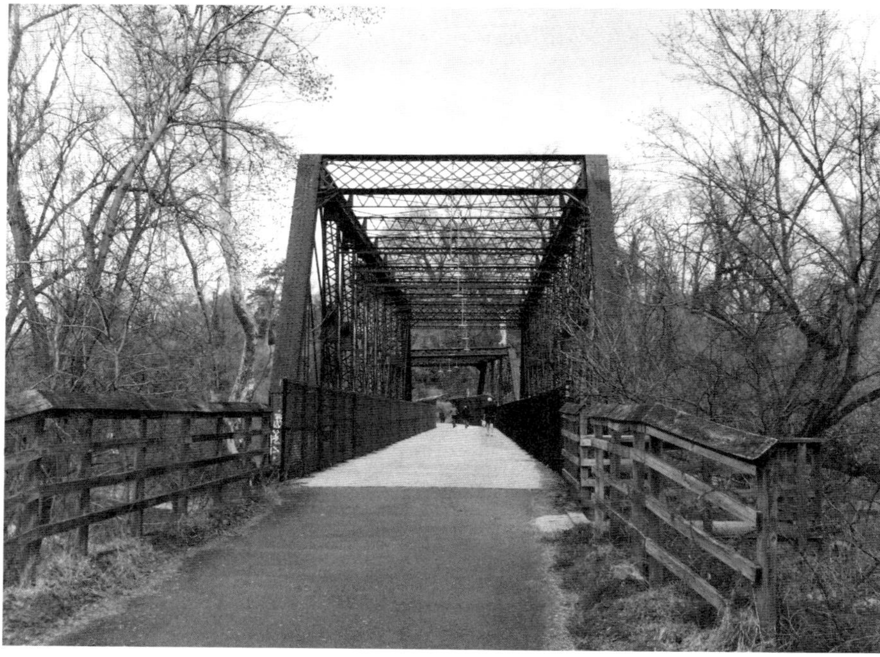

Joggers and cyclists on the Arizona Avenue Bridge, which carries the Capital Crescent Trail across the C&O Canal.

Today, the bridge carries cyclists and hikers on the Capital Crescent Trail across the canal and Canal Road. Railroad use of the bridge ceased in 1985 and a community organization, The Coalition for the Capital Crescent Trail, was created to advocate for repurposing the railroad right-of-way into a trail in 1986. By 1988, the Montgomery County Council had purchased the Maryland portion of the trail. The National Park Service purchased the segment of the railroad in the District in 1989 and 1990 and incorporated it into the C&O Canal National Historical Park. The seven-mile-long section of the asphalt rail trail from Georgetown to Bethesda opened in 1996. In addition to cyclists and hikers, the Arizona Avenue Trestle is familiar to residents of Northwest D.C. and the Maryland suburbs because Canal Road, which passes directly beneath the bridge, is a major commuting route into the city. Thousands of drivers and passengers pass underneath it every day, and it is likely that some of them notice that the black, Erector Set appearance of the bridge indicates that it is historic.

WISCONSIN AVENUE (HIGH STREET) BRIDGE

Even before the ten-by-ten-mile square of land was selected for the location of the new Federal City, the port of Georgetown was an important shipping and manufacturing center. The setting of the town at the head of navigation of the Potomac River, close to tidal water that could serve as a harbor and free-flowing fresh water that could drive mills, provided many advantages for commercial enterprises and settlement.

These advantages only accelerated in the nineteenth century as the technology was developed to construct bridges, canals, aqueducts, and railroads. Georgetown became the industrial center and transport hub of the city. It was connected to the outside world by the Little Falls Bridge in 1797 and, by 1909, was also the terminus of the C&O Canal, the trolley line to Glen Echo, and the Georgetown Branch of the B&O Railroad.

These transportation systems delivered workers, as well as raw materials, food, water, and coal to support the industrialization of Georgetown and the growth of the city. They also carried industrial products out of the city to markets across the growing nation and abroad. While much of the rest of the District developed with government office buildings and glittering white marble monuments, Georgetown brimmed with docks, warehouses, and factories.

For much of its length, the C&O Canal was located directly adjacent to the river, with little or no settlement or industrial development on the narrow strip of land between the two. However, the docks, warehouses, and factories in Georgetown required space. For the final mile of the canal between the Potomac Aqueduct and the terminus at Rock Creek, the canal diverged from the river. This allowed space for warehouses and industrial facilities on both the northern and southern sides of the canal, but also meant that bridges were needed to cross the canal.

Five of these early bridges crossing the canal in Georgetown were constructed, but only one remains today. This is the bridge at what was known, until the 1890s, as High Street, now named Wisconsin Avenue. Constructed in 1831, it is the District's oldest bridge.

The Wisconsin Avenue Bridge is a simple single-arch stone bridge constructed of Aquia Creek sandstone, one of the most widely used building stones in the District's early years. The bridge is about fifty feet long and twenty feet high over the towpath.

Even though the bridge carries two-way traffic just a block from M Street, which is one of the busiest thoroughfares and shopping districts in the city,

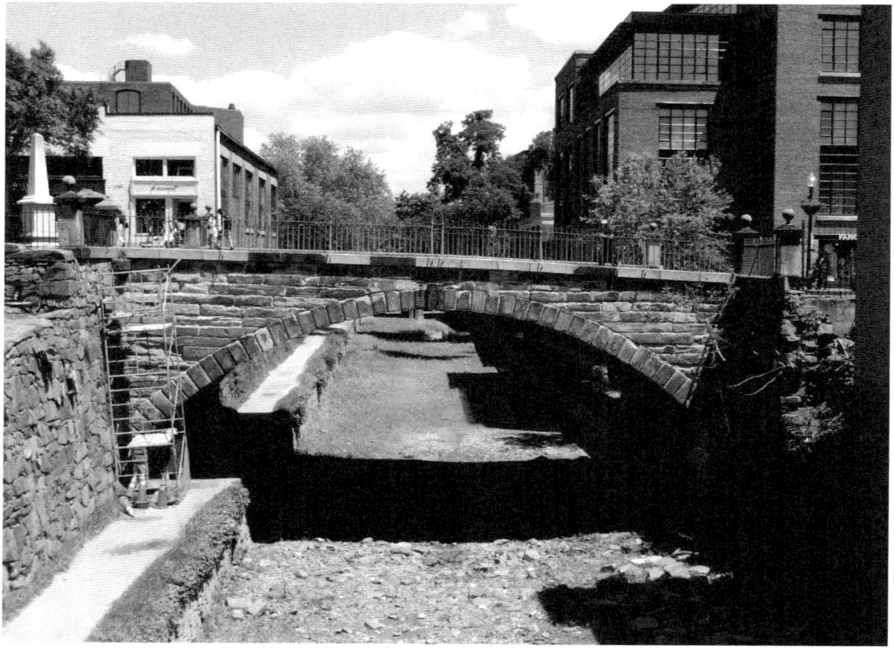

The 1831 High Street Bridge, with the 1850 obelisk placed to commemorate the completion of the C&O Canal.

its historic importance is known to few residents and even fewer tourists and visitors. This is because Wisconsin Avenue and all other parallel streets crossing the canal dead-end at Water Street, just a few blocks south of the canal. There is little car traffic in this area because the streets are not through streets, and there is very little public parking available in the narrow sliver of land south of the canal. Even pedestrians or those driving across the bridge are unlikely to notice its historic stature because it is a substructure arch bridge with no superstructure above the deck. All that is visible is a decorative green iron railing and a short obelisk, easily overlooked, that commemorates the completion of the C&O Canal.

The bridge is visible only to pedestrians walking the C&O Canal towpath, but those who do are in for a treat, because the early date and historic importance of the bridge are immediately apparent. The bridge looks old, but not in an unkempt, dilapidated way. The sandstone blocks range in color from tan to orange, with many of them blackened by decades of weathering and burning of coal. The different-sized blocks are shaped to stand out, especially those defining the arch-ring, making it visually interesting.

Most notable is that the bridge's spandrels on both sides include large blocks to display engraved inscriptions. On the bridge's western side, the block above the keystone identifies John Cox as the mayor of Georgetown and James Dunlop as the recorder. The keystone below this block is inscribed with a construction date of 1831. These inscriptions are not easy to read from the towpath, but you can read them with binoculars or a telephoto camera lens.

On the eastern side, there is a larger block for inscriptions on the spandrel on either side of the arch and a larger block used as a keystone. These were all inscribed back in the day but are largely illegible now. The bridge was renovated in 1992, giving the National Park Service a close look at these blocks, and the reported information is confusing. According to the NPS, one of the stone plaques refers to Andrew Jackson having dedicated the bridge on its completion in 1828, which is three years earlier than the date inscribed on the keystone on the western side.

The obelisk on the northwestern corner of the bridge also deserves a visit. This monument was placed on this spot in 1850 to commemorate the completion of the C&O Canal to its western terminus at Cumberland. The obelisk itself is only about five feet high but sits on a square white marble base that is also about five feet high. The base is engraved on all four sides with the names of the engineers, directors, and other officials associated with the construction of the canal.

In all, there are nine bridges crossing the canal between Key Bridge and Rock Creek, but the High Street Bridge is the only one that is historic or decorated. West of the High Street Bridge are four small footbridges connecting recent condo and restaurant developments on either side of the canal. The bridge closest to Key Bridge is named the Alexandria Aqueduct Footbridge, although it is on the opposite side of Key Bridge from the Aqueduct Arch. This footbridge links the Capital Crescent Trail and C&O Canal towpath on the southern side of the canal to the Francis Scott Key Memorial Park at the corner of M Street and Key Bridge. From the Key Memorial, cyclists and pedestrians can access the sidewalk on Key Bridge, cross the Potomac to Theodore Roosevelt Island, and connect to the Mount Vernon Trail system.

The four bridges crossing the canal between the High Street Bridge and Rock Creek are road bridges carrying, from west to east, Thirty-First Street, Thomas Jefferson Street, Thirtieth Street, and Twenty-Ninth Street. These bridges are all identical concrete deck bridges sitting on steel girders, and each carries only one lane of traffic. However, this short stretch of the very end of the canal is an interesting historical attraction on its own. In order to lower canal barges down to sea level at the mouth of Rock Creek and the Potomac River, the final half mile of the canal has four locks in short sequence. Although the street bridges are modern, each of them uses the stone walls of the historic locks as abutments. The towpath is lined with historic houses, including The Georgetown House on the northeastern corner of the Thirty-First Street Bridge, constructed in 1830 to support canal operations. The Georgetown Visitor Center for the C&O Canal National Historical Park is also located here, on the northeastern corner of the Thomas Jefferson Street Bridge, across from Lock 3.

5

ANACOSTIA RIVER

In the late twentieth century, the name Anacostia had a negative association for the residents in many parts of the District and its suburbs. When the city was founded in 1790, the Eastern Branch was shallow, bordered on both shores by tidal mudflats, and only navigable for a short distance inland. A 1912 article in the *Sunday Star* directs readers to visit the location of the original Eastern Branch Bridge by referring to the "weedy, slimy, oozey" shoreline of the Anacostia.

George Washington saw a major advantage in siting the Federal City at the confluence of the Potomac and its Eastern Branch. The Continental navy was dismantled following the Revolutionary War, but Washington foresaw that a navy might one day be needed. The Eastern Branch offered river frontage that could be used for shipbuilding for a national navy. An area on the northern shore of the Eastern Branch was set aside for this purpose, and the Washington Navy Yard began constructing ships to support the Tripolitan War in 1802. Thomas Jefferson substantially expanded the operations of the Navy Yard, and it rapidly became the second-largest employer in the city, after Congress. Other similar developments followed, establishing the industrial character of the Anacostia River for the next two hundred years.

This industrialization happened when there was neither environmental regulation of waste disposal or sewage discharge nor protection of wetlands or fisheries. Therefore, the industrialization brought infilling of the tidal wetlands along the shore as well as chemical and sediment pollution, both of which largely destroyed the natural ecological function of the river. As a result, the Anacostia River became the unseen, less desirable river in the city, a place where no one would want to go for tourism or recreation. For those who grew up in the region in the late twentieth century, the Potomac was the monumental river, while the Anacostia was the industrial river.

Racial segregation in what was very much a Southern city contributed to the lack of tourist or recreational attractions along the Anacostia. Maryland and Virginia permitted slave ownership, and much of the construction of the original road network and federal buildings, including the Capitol and the President's House, was completed using enslaved labor. The location of the city south of the Mason–Dixon Line, instead of in the established Northern cities of New York or Philadelphia, was largely a compromise with the slaveholding Southern states. The city was managed by commissioners, many of whom were slaveholders, as were ten of the first twelve presidents, who appointed the commissioners.

This racially based origin and urban development led to segregation of white and Black neighborhoods in the city, along with all of the inferior infrastructure, poor schools, and lack of economic opportunity that segregation causes. The Anacostia River, as the prominent natural boundary within the city limits, became a dividing line. A large influx of freed Blacks migrated to the city in the years following the Civil War, and many settled in the Anacostia neighborhood south of the river of the same name. In the late 1950s, the development of the elevated Anacostia Freeway along the southern shore of the river made sure there was no visual connection between Anacostia and the rest of the city.

As recently as twenty years ago, few people from outside of Anacostia would choose to go there for recreation or tourism. Because Anacostia was separated from the monumental core of the city, few museums, monuments, or other prominent tourist attractions were developed south of the river. In addition, the dead, polluted river adjacent to a segregated, economically depressed neighborhood provided no attraction for parks or recreation facilities. The few bridges that were there did have sidewalks to allow Anacostia residents to walk across the river to their jobs, but their primary purpose was to allow automobiles driven by suburbanites to access the city center via the Anacostia Freeway.

Today, the situation is far different. Older residents in the suburbs still recoil at the suggestion that there is any reason to go to the other side of the Anacostia, but that is only because they have not been there to see the remarkable changes. Metro stations east of the river on the Blue and Orange Lines were opened in 1980, and the Green Line through Old Anacostia opened in 1991. This greatly improved access to the center of the city for commuters and increased the desirability and economic conditions of these neighborhoods.

A similar transformation occurred with respect to the ecology of the river. Cleanup efforts on the Anacostia began in the 1980s and continue today. There are so many different nongovernmental organizations and interagency committees working on various improvements that it is impossible to keep track of them all. The Anacostia Watershed Society was formed in 1989, with the goal of cleaning the river to the extent that fishing and swimming could once more be allowed. A major step in this direction was a 2006 Supreme Court ruling that determined that sewage dumping violated the Clean Water Act, effectively stopping this practice. Industrial activities at the Navy Yard ceased in the 1960s, and massive volunteer efforts have been mobilized to clean garbage and waste from the river banks.

Concurrent with these improvements in public transportation and environmental conditions, the District's government, the adjacent county government of Prince George's County, the National Park Service, and several other federal and nonfederal partners formed the Urban Waters Federal Partnership to establish parks and multiuse trails on both sides of the Anacostia. In fewer than twenty years, the shores lining both sides of the entire length of the Anacostia River have been transformed into a continuous network of parks providing a remarkable variety of recreational opportunities and community gathering spots. These parks are in walking distance and available to residents in neighborhoods on both sides of the river. Removal of the industrial facilities has opened the northern shoreline to the development of riverfront condos and office complexes, and new residential developments are also springing up in Anacostia near the Frederick Douglass Bridge. Integral to all of this development was improvement of the walkability of the bridges by including wider sidewalks and easy connections from the bridge sidewalks to the trails passing below.

BLADENSBURG WATERFRONT AREA

The Bladensburg Road Bridge and the Bladensburg Park Pedestrian Bridge at the Bladensburg Waterfront Park are in Maryland and not within the District's boundaries. However, they are located only about one mile north of the Maryland-District boundary near the confluence of the Northwest and Northeast Branches, and therefore including them in a bike tour allows you to complete the entire Anacostia River Trail system. This is not just for people who hate leaving things incomplete, but rather it is an integral part of the overall rebirth of the parks and trails along the Anacostia River.

Neither bridge is historic, decorated, or of unusual engineering, and neither would make the list of bridges to visit by most bridge enthusiasts. Although the Bladensburg Road Bridge does have sidewalks, there is no ramp or staircase connecting the sidewalks to the Anacostia River Trail, which passes beneath the western end of the bridge.

The pedestrian bridge, however, does serve a critical function in connecting the entire Anacostia Trail system to the regional trails in the Maryland suburbs. Because of a complex traffic interchange on the eastern end of the Bladensburg Road Bridge, the trail must be on the western side of the river as it flows south of the confluence. However, other developments on the western side of the river, including the U.S. National Arboretum and the Langston Golf Course, preclude a continuous trail on the western side.

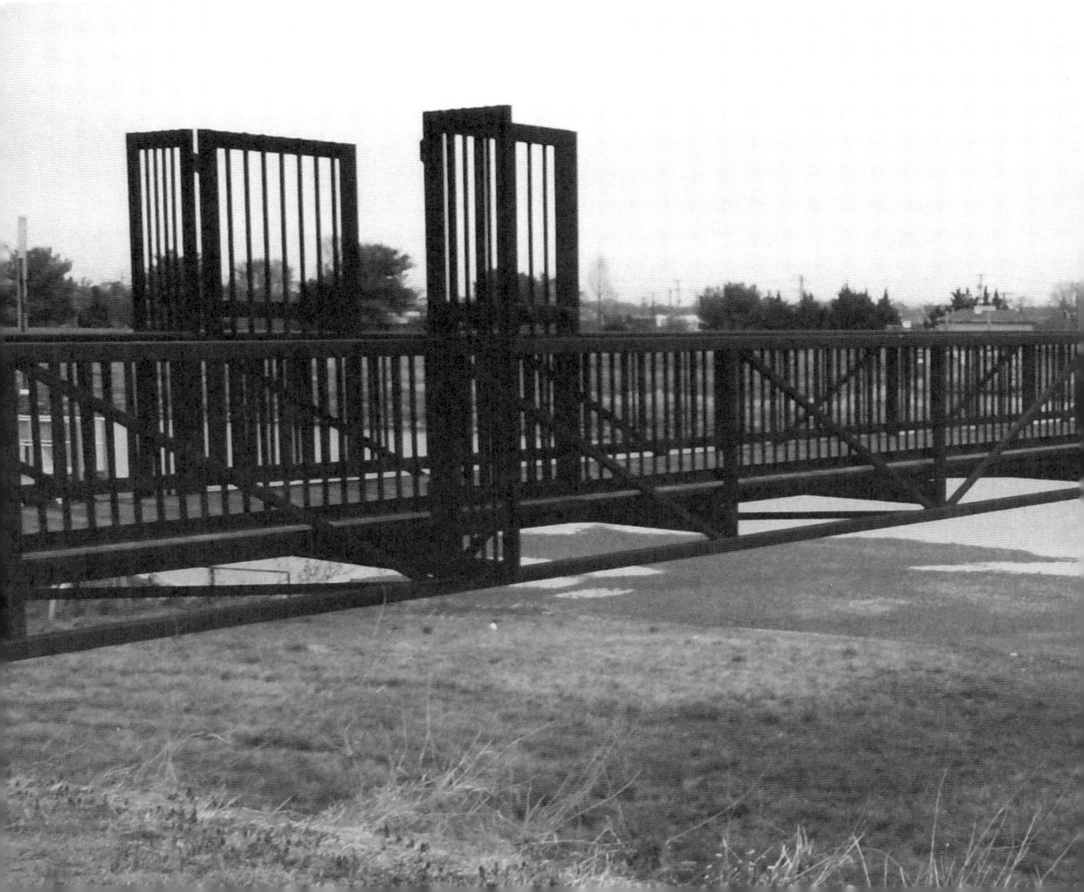

Therefore, a pedestrian bridge crossing the river at Bladensburg is used to carry the trail from the western side to the eastern.

The Bladensburg Park Bridge was constructed in 2006 and consists of two spans crossing the river and the adjacent marsh. The main span over the river is supported by a superstructure arch made of cor-ten steel, and the part of the bridge over the marsh is a flat steel-girder span. The two spans are at different orientations, so there is an angle in the middle of the bridge that, along with the superstructure arch, gives the bridge some visual appeal. The bridge is about ten feet wide, four hundred feet long, and fifteen feet above the river. No prominent tourist sites are visible from the bridge, although waterbirds such as cormorants can be seen resting on the tops of abandoned pylons of a former dock in the marsh.

The bridge carries the Anacostia Trail through the Bladensburg Waterfront Park, which is operated by the Maryland-National Capital

View of bridge at Bladensburg connecting the Maryland bike trail system to the Anacostia bike trail system.

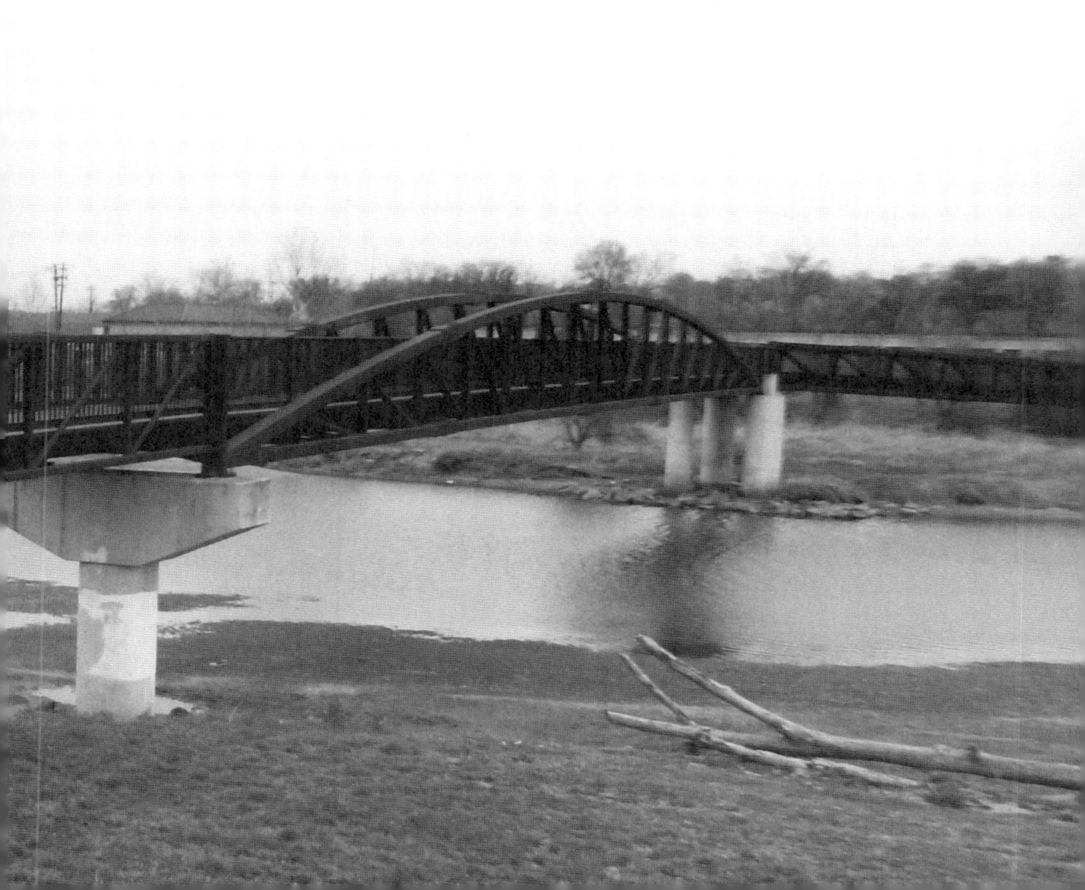

Park and Planning Commission. In addition to the trail, the park has a boat ramp, fishing pier, benches, gazebos, restrooms, covered picnic tables, ample parking, and other general park amenities. The park is home to the Washington Rowing School, which offers rowing courses on the river. Numerous placards describe the history and ecology of Bladensburg, as well as a map inscribed into a concrete plaza depicting the original layout of the town lots in 1742.

Although the park and trail system along the entire length of the Anacostia is reason enough to recommend venturing a bit outside the District's boundaries to visit the Bladensburg Waterfront, the more important reason is the role of the original bridge at Bladensburg in the history of the District.

At the founding of the Federal City in the 1790s, most of the population, industry, and finance were situated to the northeast in Baltimore, Philadelphia, New York, and Boston. As a result, most of the foot, horseback, and carriage traffic in and out of the new capital city came through the Baltimore Pike, creating the need for an early bridge carrying this road over the Eastern Branch. That bridge crossed at Bladensburg, close to the current location of the Bladensburg Road Bridge.

Information on the construction date of that bridge is scarce. In 1976, a diorama of the port of Bladensburg as it existed in 1770 was commissioned by the Prince George's County Jaycees, and it depicted a wooden bridge. A map was drawn of the city in 1782 by one of the engineers of the Comte de Rochambeau, who camped there on his way to Yorktown. That map also shows a bridge. These imply that a bridge connecting Baltimore to what is now the District was in place more than twenty years before the location was selected as the site of the new Federal City.

The original Bladensburg Bridge was not historically important just because it connected the Federal City to the Northern cities. More critically, the bridge was the focus of the Battle of Bladensburg in the War of 1812, making it the most consequential bridge affecting the history of the city, and possibly the nation.

In 1814, after the Americans destroyed the capital of York (now Toronto) in Upper Canada, the British invaded the Chesapeake Bay, landing their troops at Benedict on the Patuxent River southeast of the District. On August 24, 1814, the Americans massed their troops to prevent a crossing at the Eastern Branch Bridge near the Navy Yard. The British, however, went overland to the north to Bladensburg, where there was a ford and a wooden bridge approximately 120 feet long and twelve feet wide. The Americans hastily repositioned their troops in reaction to the attack from this unexpected direction, but they were too late. The British crossed the bridge, defeating the Americans in a battle a short distance west of the river at the current location of Fort Lincoln Cemetery. After entering Washington and burning both the White House and the Capitol, the British retreated back over the bridge and proceeded to attempt an attack on Baltimore that inspired District resident Francis Scott Key to pen "The Star-Spangled Banner."

Although there is no battleground area or visitor center available for interpretive tourism in relation to the battle, you can visit multiple sites in the area. Historic information placards in Waterfront Park, as well as along roadsides in what is now Colmar Manor on the western side of the river, describe the troop movements of the battle. Finally, the staff at the small headquarters building includes a naturalist and a historian who lead boat tours and history lectures and are more than happy to provide information on the battle.

Although the park and bridges can be visited on foot or bicycle, a more thorough visit to understand the various battle sites requires a car. Boat tours and lectures discussing the history of the area are centered at the park, where there are large parking lots.

ETHEL KENNEDY AND WHITNEY YOUNG BRIDGE AREA

Although the Ethel Kennedy, Whitney Young, and Heritage/Kingman Island Bridges are all very different types, with different histories and purposes, the discussion of the recreational and tourist uses of these bridges is combined into a single unit. The bridges are separated by only half a mile and are linked together by three separate park and trail systems. As a result, cyclists and pedestrians commonly enjoy them together, as a group.

The location of what is now called the Ethel Kennedy Bridge, formerly known as the Benning Road Bridge, is one of the more historic bridge locations in the District. A bridge at East Capitol Street was envisioned in 1791 by L'Enfant but was not constructed at that time. Instead, the Maryland Assembly gave a charter to the Anacostia Company in 1797 for a bridge about a half mile upstream. Although it was chartered after the Lower Bridge, this Upper Bridge was constructed first, in 1800. The reason for the location farther north of East Capitol Street was because Benjamin Stoddert, who owned lands east of the river and wanted to develop them, provided the funds for bridge construction, resulting in the name of Stoddert's Bridge. Later, the shares of the Anacostia Company passed to Stoddert's son-in-law Thomas Ewell, leading to references as Ewell's Bridge.

The Upper/Stoddert's/Ewell's Bridge was burned by the Americans on August 24, 1814, to stop the British from entering the city. It was rebuilt in 1820 and then auctioned to William Benning in 1823. Benning reconstructed the bridge in 1830, and it was eventually replaced by an iron bridge in 1892, a concrete-encased steel structure in 1934, and the current steel girder bridge carrying Benning Road in 2004. In 2014, the bridge was renamed for Ethel Kennedy, a prominent human rights and environmental protection activist and the widow of Robert F. Kennedy.

The Ethel Kennedy Bridge marks a change in the management and character of the land on the shores of the Anacostia. North of this bridge, there is only one Anacostia River Trail, which is on the eastern shore of the river. From this point south to the Frederick Douglass Bridge, a distance of approximately two and a half miles, there are continuous parks and riverfront trails on both sides of the river. The sidewalk on the Ethel Kennedy Bridge provides a pedestrian and bike connection between the East Bank Trail and the West Bank Trail and also has a gate offering pedestrian access to the hiking and nature trails on Heritage and Kingman Islands.

Although a bridge carrying East Capitol Street was proposed by L'Enfant in 1791, the Whitney Young Bridge is one of the most recent bridge locations in the city. Constructed in 1955, it is a low-profile, midcentury steel-girder bridge sitting on stone-faced piers. The bridge carries two lanes of traffic in each direction, with a sidewalk on the northern side, facing Heritage and Kingman Islands. On the eastern end, the bridge continues the straight route of East Capitol Street further into the Fort Dupont and Capitol View neighborhoods of eastern Washington, D.C. On the western end, the roadway enters a complex of circles around RFK Stadium, the former home of the Washington Senators and Nationals baseball teams, the Washington Redskins football team, and the D.C. United soccer team, before continuing directly west to the Capitol. The name of the bridge was changed in 1973 from East Capitol Street Bridge to the Whitney Young Memorial Bridge in honor of the civil rights activist and former executive director of the National Urban League.

The recreational link between the Ethel Kennedy and Whitney Young Bridges consists of a variety of parks and trails on both banks of the Anacostia and even on Kingman Island in the middle of the Anacostia. On the eastern side, the East Bank Trail passes through River Terrace Park, which is situated between a residential development and the river. River Terrace Park is a unit of the National Capital Parks–East. The park occupies the eastern shore of the Anacostia between the Ethel Kennedy and Whitney Young Bridges. You can use the East Bank Trail to access either of the bridges.

On the western side of the river, the city opened The Fields at RFK Campus, operated by the city through Events DC, the city's convention and sports authority, in 2019. The campus occupies what were previously asphalt parking lots associated with RFK Stadium and, in addition to the West Bank Trail, includes three full-scale soccer fields and other amenities.

In the river between these two parks, two artificial islands were created through disposal of dredge spoils to form the Burnham Barrier in the early twentieth century. In 2018, Kingman and Heritage Islands were designated as a city park for ecological conservation and education, operated by the D.C. Department of Energy and the Environment.

There are two access points for the nature trails on Kingman and Heritage Islands. On Ethel Kennedy Bridge, the eastern segment crosses the main channel of the Anacostia, and the western segment crosses over a narrow channel of Kingman Lake. In between, Benning Road crosses Kingman Island, which is the artificial island that separates Kingman Lake from the Anacostia. The sidewalk on the southern side of the bridge is lined with a black steel fence. About halfway across, where the bridge crosses the island, a gate in the fence provides pedestrian and bicycle access to Kingman and Heritage Islands Park.

Alternatively, the islands can be accessed using an entrance at the parking lots for The Fields at RFK Campus. Signs identify the entrance, which is just off the West Bank Trail. The paved entrance trail passes through a gate in a black steel fence and is marked with a wooden sign with carved figures of a heron, dragonfly, and butterflies, indicating that it is an environmental preserve. Just inside the gate, there are benches, information placards, and an interactive kiosk providing weather, tide, and other data to support recreational fishing and boating.

The entrance trail then crosses over a small wooden footbridge onto Heritage Island. At the eastern end of the bridge, the trail becomes an unpaved nature trail, which completes a circuit of Heritage Island. Heritage Island is entirely enclosed within Kingman Lake. However, another wooden footbridge connects the trails on Heritage Island to the nature trails on the larger Kingman Island to the east. This bridge has two different sections connected to the islands and a large wooden platform in the middle. The

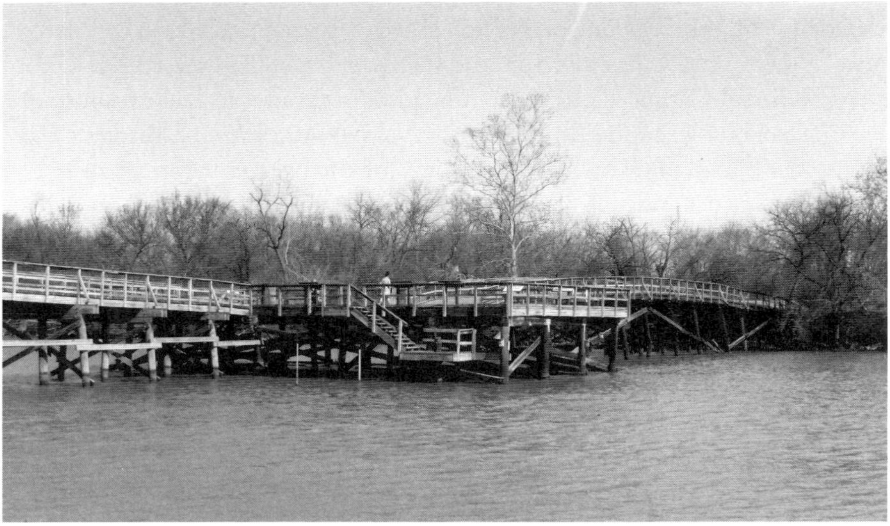

Bridges and fishing platform connecting Heritage Island to Kingman Island across Kingman Lake on the Anacostia.

large platform has more benches and ecological information placards, as well as stairs leading down to platforms that can be used for fishing or access to canoes and kayaks.

On Kingman Island, the nature trail continues about a half mile south, crossing underneath the Whitney Young Bridge, but with no access from the bridge to the island. It also leads about a quarter mile to the north, to the gate in the middle of Ethel Kennedy Bridge.

JOHN PHILIP SOUSA BRIDGE

The 1791 L'Enfant Plan's layout for the Federal City envisioned two bridges across the Eastern Branch. One was a continuation of East Capitol Street directly east, and the other was a continuation of Pennsylvania Avenue across the river to the southeast. These locations are currently occupied by the Whitney Young Bridge carrying East Capitol Street and the John Philip Sousa Bridge carrying Pennsylvania Avenue. Of the two, the bridge at East Capitol Street was not constructed until 150 years after it was first proposed. However, an early bridge was constructed close to the current location of the Sousa Bridge.

The Eastern Branch Bridge Company was chartered by the Maryland Assembly in 1795 to construct a bridge that would connect the lower end of Kentucky Avenue north of the river to Matthew Wigfield's farm south of the river. Kentucky Avenue intersects Pennsylvania Avenue close to the Anacostia shoreline, so this bridge was approximately in the location intended in the L'Enfant Plan. The location had been the site of a ferry, variously referred to as the "Wigfield Ferry," "Naylor-Talbert Ferry," and "Dyer's Ferry." The charter described a wooden bridge twenty feet wide with a drawspan sufficient to allow passage of a thirty-foot-wide ship. In *Bridges and the City of Washington*, Donald Beekman Myer describes the construction date of the bridge as 1804, while Van Dyne's 2006 article in *The Washingtonian* reports it to be shortly after 1800.

There were not only multiple names for this bridge, but one of the names used was later transferred to a different bridge as well. Most sources refer to it as the Eastern Branch Bridge, while others call it the Anacostia Bridge, the Navy Yard Bridge (because of its proximity to the Navy Yard), and the "Lower" Bridge (to distinguish it from the Upper Bridge farther upstream).

For decades, it was commonly assumed that the British burned the bridge during their attack on Washington, D.C., on August 24, 1814. However, it was later discovered that the Americans had destroyed the bridge in an attempt to stop the British from entering the city. In the 1880s, a realtor named M.I. Weller, who was conducting historical research as part of his effort to lobby Congress for a new bridge, found a firsthand account in the National Archives. The report was written by a Captain Creighton, who related that he had carried out orders from Commodore Tingey of the U.S. Navy to destroy the bridge with gunpowder and by lighting it on fire. Captain Creighton also destroyed the Upper/Stoddert's Bridge as well.

The Eastern Branch Bridge was quickly rebuilt, but it was apparently not well-maintained. A newer bridge, situated closer to the Navy Yard and providing easier access between the Navy Yard and the new suburb south of the river called Uniontown, was constructed at Eleventh Street in 1820. The names of "Navy Yard Bridge" and "Lower Bridge" were transferred to this new bridge at Eleventh Street, and the original bridge at Pennsylvania Avenue came to be called the Middle Bridge.

In 1845, sparks from a passing riverboat lit the bridge on fire, and it was not rebuilt until a new bridge, with a slightly different orientation, was constructed at Pennsylvania Avenue in 1890. The 1890 bridge was a steel-deck truss with ten spans and two through-truss spans crossing the Baltimore and Potomac Railroad line on the northern shore.

Eventually, an increase in traffic rendered the bridge obsolete. Construction on the current bridge began in 1938, and it was completed in 1940. The bridge is a steel-plate girder bridge, consisting of nine spans on concrete piers that are faced with stone. The bridge, constructed with thirty-six feet of clearance, did not require a drawspan.

In 1939, Congress named the bridge after John Philip Sousa. The famous composer was a Washington native, born in Southeast D.C., and buried in the historic Congressional Cemetery not far from the end of the bridge.

Although the bridge was one of several midcentury steel-plate girder highway bridges constructed in the city, it is more attractive and recreation-friendly than most of the other bridges of the same time frame. The main aesthetic difference, when viewed from the side, is subtle. The steel-plate girders on the Sousa Bridge are curved, giving the bridge profile a pleasant scalloped appearance. Another difference, again subtle, is that the Sousa Bridge is a straight continuation of Pennsylvania Avenue from the Capitol across the river and southeast into Maryland. This aligns with the geometry of the 1791 L'Enfant Plan. The simplicity of the alignment of the Sousa Bridge, along with ramps connecting both sidewalks to the East Bank and West Bank Trails that pass underneath, integrates the Sousa Bridge into its setting.

Another pleasant feature is that the sidewalks are lined with attractive steel railings and tall lampposts with geometric designs. At the ends, the top

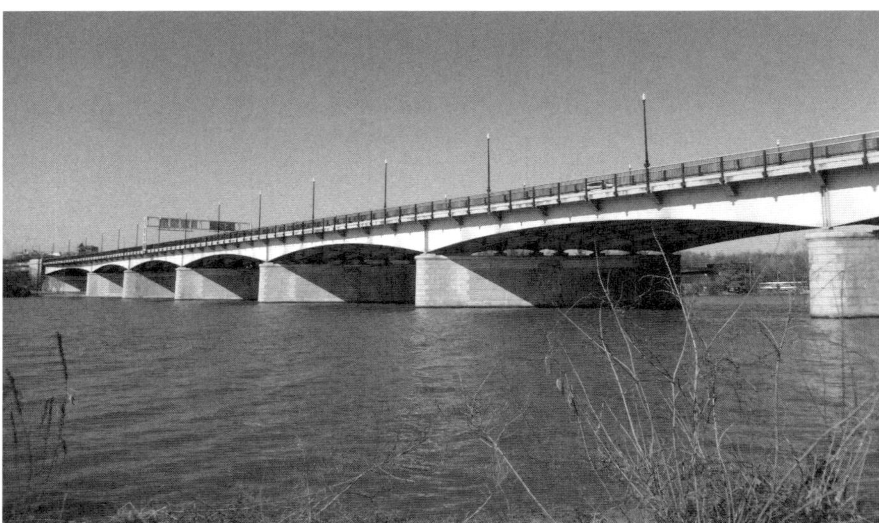

The John Philip Sousa Bridge, constructed across the Anacostia in 1940.

of the piers along the sidewalk are made of carved granite, with benches and decorative name and date plaques engraved into the stone.

The northern landing of the bridge crosses the West Bank Trail and CSX rail lines and then passes through Barney Circle, named for Joshua Barney, a Revolutionary War veteran who played a prominent role for the U.S. Navy in the War of 1812. From there, Pennsylvania Avenue leads directly through a residential neighborhood and to the Capitol. On the southern landing, the shore of the Anacostia in both directions is Anacostia Park, consisting of grassy lawn areas, the East Bank Trail, and the locally famous Pirate Ship Playground.

ELEVENTH STREET (NAVY YARD) BRIDGE

The Eleventh Street Bridge is a complex of three bridges that connects the Anacostia Freeway to Interstate 695 leading into the city center. From a distance, the bridges look similar to the Fourteenth Street Bridge complex, appearing to be nothing more than large-scale, utilitarian, uninteresting bridges designed for nothing more than carrying large volumes of commuters as quickly as possible. However, that impression belies the fact that it is one of the most historic bridge locations in the city, having played a prominent role in national historic events. Today, it is a central part of the renaissance of community-focused bridge construction, designed to serve the District's residents for years to come.

The bridge is located about a mile west of the Sousa Bridge at Pennsylvania Avenue. Some of the documentation on the history of bridges at the two different locations contains errors and creates confusion. The earliest bridge at these locations was the 1804 Eastern Branch Bridge at Pennsylvania Avenue, which was also known as the Lower Bridge or the Navy Yard Bridge. This was the bridge that, along with the Stoddert's/Upper Bridge, was partially burned by the Americans, not the British, on August 24, 1814. The Lower Bridge was rebuilt shortly after but burned in 1845.

At Eleventh Street a mile away, a separate bridge was constructed by 1820. Because this newer bridge was closer to the Navy Yard and "lower" on the river, these monikers that had been used for the older bridge at Pennsylvania Avenue became applied to the newer bridge at Eleventh Street. Because the name was transferred—and the actual story of the burning of the Lower

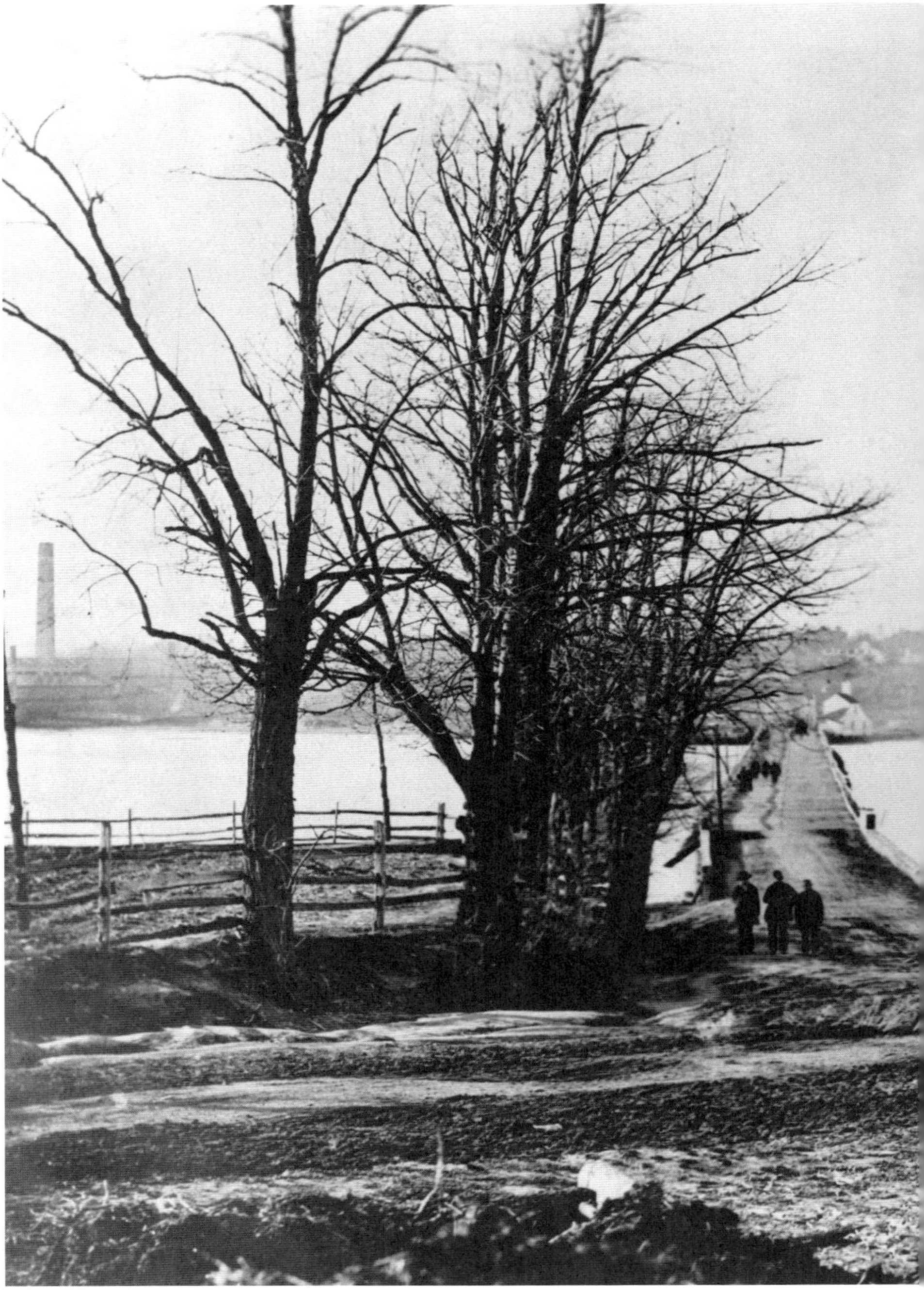

The Eastern Branch Bridge, also known as the Lower Bridge and Navy Yard Bridge. *Library of Congress.*

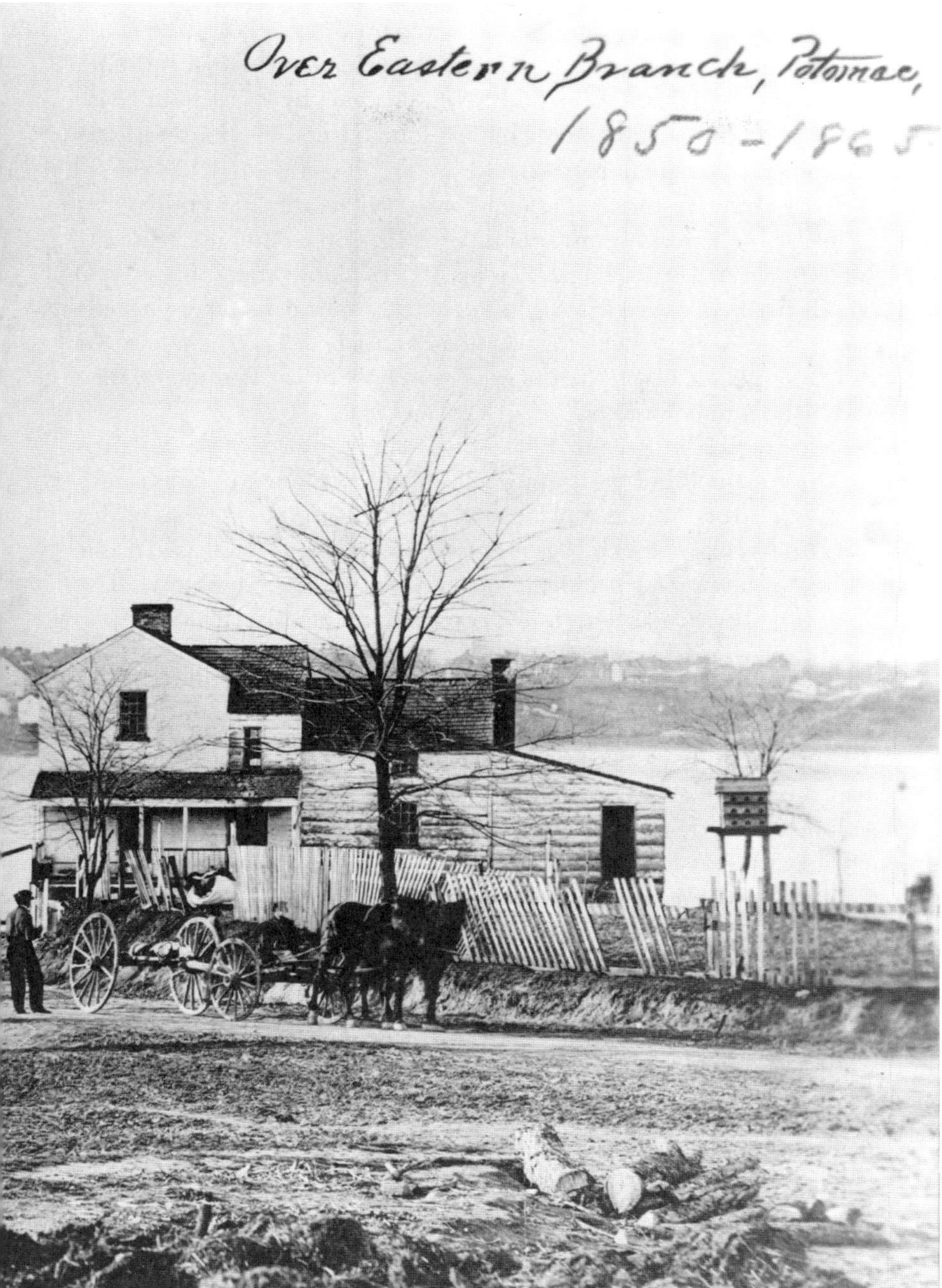
Over Eastern Branch, Potomac,
1850-1865

Bridge by Captain Creighton was not discovered until later in the nineteenth century—multiple sources report that it was the original bridge at Eleventh Street that was burned by the British in 1814 and again by accident in 1845. It was actually the bridge at Pennsylvania Avenue burned by the Americans in 1814 and then by accident in 1845.

The one accurate historical highlight ascribed to the Eleventh Street Bridge is its role in American history on April 14, 1865. After assassinating Abraham Lincoln at Ford's Theatre, John Wilkes Booth made his way to the Eleventh Street Bridge, where he apparently talked his way past the army guards, who were unaware of what had happened. Once south of the Anacostia, Booth made his way to the northern shore of the Potomac, which he crossed by boat into Virginia.

The history of the bridges at Eleventh Street follows a pattern that is similar to that of the other early Anacostia bridges. The early nineteenth-century wooden structure was replaced by a wrought-iron deck-truss bridge in 1874, and that replacement bridge became obsolete due to increased traffic and invention of automobiles and trucks in the early twentieth century. The 1908 replacement was a steel-arch bridge with a double bascule drawspan and carried trolleys as well as traffic.

After the massive increase in traffic levels due to the new Anacostia Freeway in the 1950s, the 1908 bridge was replaced in the late 1960s by utilitarian twin interstate highway bridges similar to those constructed at Fourteenth Street over the Potomac. The twin bridges served their purpose by moving larger volumes of commuter traffic for decades, but they also visually separated the Navy Yard from the river and the Anacostia neighborhood. A 1974 summary of Washington's bridges by Donald Beekman Myer of the U.S. Commission of Fine Arts bemoaned the bridge as having random unmatched components and no relation to the historic street pattern, claiming Eleventh Street was now "lost."

A review of the District's highway system shows that almost all of the Potomac and Anacostia bridges carry local surface streets as opposed to high-speed interstate highways. Many, maybe even all, of these bridges are jam-packed with commuters during rush hour, trying to access either the George Washington Memorial Parkway or Anacostia Freeway. Only three bridges carry high-speed interstate highway traffic into the city, and the Eleventh Street Bridge is the only one of those that connects to the Anacostia Freeway. This resulted in the twin bridges at Eleventh Street being one of the most heavily used bridge complexes in the city. By the early 2000s, their design capacity had been exceeded.

Starting in 2002, the District Department of Transportation (DDOT) began conducting structural assessments of the bridges, and a decision to replace them was reached in 2004. But this bridge replacement project was far different from others in the District, and it is important to consider what else was happening at the same time.

By the early 2000s, the renaissance of incorporating aesthetics and community-friendly features had been influencing bridge design in other parts of the nation for ten or more years. The District was still dominated by unattractive, utilitarian highway bridges. However, the 2004 Benning Road Bridge, which was the first major new bridge in the city since 1972, had adopted some enhancements. Although still a utilitarian bridge, it improved on the midcentury highway bridge design by including wide sidewalks and a white concrete facing, which was far more attractive than the rusting steel-plate girders on other bridges. At the same time, the paradigm-altering Woodrow Wilson Bridge was already under construction south of the city.

Most importantly, the city and federal governments had begun a cleanup of the Anacostia River, including the redevelopment of the Anacostia shoreline and the adjacent neighborhoods. The decision to construct a new baseball stadium at Nationals Park on the western side of the Navy Yard completely changed the future of the neighborhood on the bridge's northern end. No longer would the Eleventh Street Bridge serve only to funnel high-speed traffic from the Anacostia Freeway to Capitol Hill, bypassing industrial and disadvantaged neighborhoods on both shores. Instead, the replacement of the twin bridges at Eleventh Street resolved this issue by constructing not two replacement bridges, but three.

Two of the replacement bridges are utilitarian, interstate highway bridges, still carrying high-speed commuter traffic from the Anacostia Freeway to Capitol Hill. What is different is that a third bridge was added in 2013, and a fourth bridge is still being planned.

The third bridge, on the western side of the twin interstate bridges, is for local traffic only, connecting the Anacostia neighborhood to the new developments surrounding the Navy Yard and Nationals Park. It connects the street network in Old Anacostia directly to Eleventh Street along the eastern boundary of the Navy Yard. Because it does not have any connections to the twin-interstate bridges and serves only local traffic, it is not jammed with commuters from the Anacostia Freeway.

The third bridge is also pedestrian and bike-friendly, but in a few different ways. It has a wide sidewalk on the southwestern side and offers direct connections to the East and West Bank Trails on either end. However, this

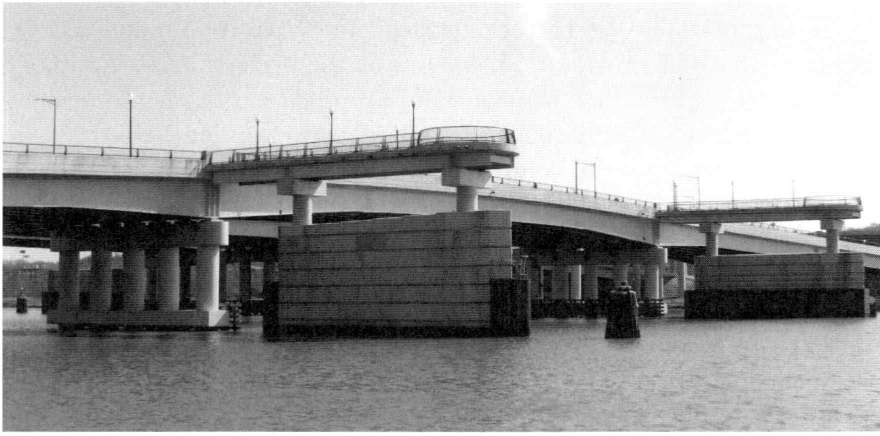

The piers of the old Eleventh Street Bridge being used, temporarily, as observation platforms.

sidewalk is just the first of many steps toward turning the Eleventh Street Bridge into a fully community-focused bridge.

The next step, which is not obvious unless you walk or bike across to examine the details, is that the piers of one of the 1960s twin bridges are still in place, only about ten feet southwest of the new sidewalk. Because they sit only about ten feet away, observation platforms have been added on top of the old piers, and these are connected to the sidewalk of the new bridge. Each platform is about fifty feet long and ten feet wide. Benches are provided along with ornate lampposts. The vista from the platforms is spectacular, giving an unobstructed view of the new Frederick Douglass Bridge and the glittering new developments of the shoreline west of the Navy Yard. These piers were not just left in place to avoid the cost of their removal. The platforms serve as placeholders, a promise of what is to come. See chapter 9 for future plans for the Eleventh Street Bridge.

YARDS PARK FOOTBRIDGE

The Yards Park Footbridge is not an important piece of transportation infrastructure, nor does it cross the Anacostia River or any other substantial body of water. However, it serves as a prominent work of public art in a vibrant, newly redeveloped part of the city and is a critical link in a continuous waterfront bike trail that stretches from Nationals Park to Kingman Island.

One of the major challenges to completing a continuous network of riverfront trails on the northern bank of the Anacostia was the presence of the Navy Yard. For more than 160 years, the Navy Yard was one of the District's biggest industrial employers. In addition to ships, it was the largest manufacturer of naval ordnance in the world. All of these activities required river frontage, precluding not only riverfront recreational trails but also any waterfront commercial or residential development. Industrial operations ceased in the 1960s, but the Navy Yard continued to house the administrative headquarters and historical museums for the U.S. Navy. The historical museums included displays of naval vessels on the former docks of the shipyard, again limiting other uses of the waterfront.

Other industrial developments also occupied the waterfront west of the Navy Yard. In the nineteenth century, the Washington City Canal connected Tiber Creek, near the Capitol, to the Anacostia, exiting just west of the Navy Yard. Also, the District Water and Sewer Authority, known as DC Water, has operated the ornate O Street Pumping Station on the waterfront since the early twentieth century.

In the early 2000s, the city and federal governments implemented a business improvement plan for the Capitol Riverfront neighborhood, also known as the "Near Southeast." One of the most visible components of this urban redevelopment plan was Nationals Park, but it also included redevelopment of the western part of the Navy Yard into the headquarters for the U.S. Department of Transportation, as well as riverfront condominiums and office buildings. The result was a remarkable complex of sparkling office and condominium buildings, sidewalk cafés, urban park features, public art, and footbridges connecting the individual pieces into an integrated urban waterfront playground.

Redeveloping this area into a continuous, accessible waterfront park required a coordinated and cooperative effort on the part of all occupants of the waterfront. Most of the privately owned properties were purchased and demolished to make way for new condos and restaurants. Although the administrative offices of the U.S. Navy remained at the Navy Yard, the former activities that required direct access to the river, including tourist visits to naval ships, were eliminated to provide public access on a narrow ribbon of waterfront between the main facility and the river. DC Water provided the same access for development of the area between the O Street Pumping Station and the river.

The northern end of the Eleventh Street Bridge lands at the eastern entrance to this riverfront quay on the Navy Yard's property. The twenty-foot-wide quay is bordered on the landward side by a wall enclosing the active Navy Yard facility. Although the pedestrian West Bank Trail along the Anacostia is continuous, the U.S. Navy prohibits bicycles on this promenade, so cyclists must walk their bikes through this short, quarter-mile-long section. Within this area, a narrow, moveable galvanized steel plank bridge can be lifted to allow access for small boats from the river into the administrative complex.

Upon exiting the Navy Yard, the promenade continues west into The Yards Park, which occupies five acres of waterfront between the Navy Yard and the O Street Pumping Station. Opened in 2010, the park is operated by Capitol Riverfront Business Improvement District (BID), a nonprofit organization. More than a promenade, The Yards Park includes kayak docks, lawn areas and a dog park, and water features.

The water features were constructed near the former channel of the Washington City Canal. The area now includes dancing fountains, which are lit at night and available for residents to cool off on hot summer days. Most importantly, the water feature is crossed by a pedestrian bridge, which

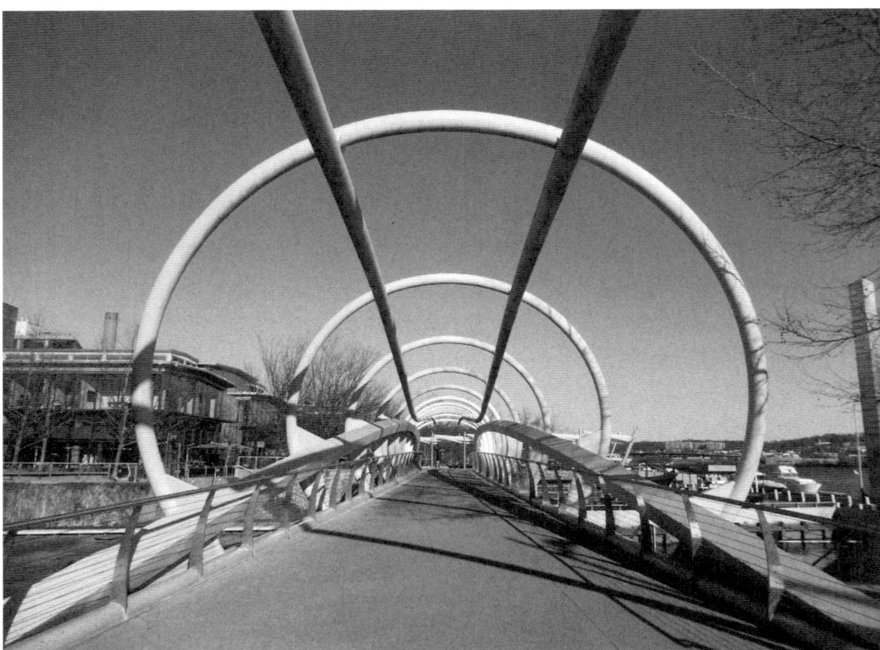

Axial view along the deck of the Yards Park Bridge on the Anacostia Waterfront.

carries bike traffic up and over the busy pedestrian promenade and serves as a modernistic work of public sculpture.

The footbridge takes the form of a two-hundred-foot-long tube made of white tubular steel. The tube is about twenty feet in diameter at each end but narrows in the middle before it opens up again. At night, the interior is illuminated by bright white lighting. Because the tube narrows in the middle, it gives an optical illusion of being much longer than it is, rendering the impression of a modernistic portal to a different dimension. This modernistic white tube is seen against the background of the more colorfully lit arches of the nearby Frederick Douglass Bridge.

Just west of The Yards Park footbridge, the trail crosses another pedestrian bridge in front of the O Street Pumping Station. Since the pumping station still functions, the stormwater discharge outlet along the riverfront could not be filled in. To allow for continuation of the Anacostia Riverwalk trail, an aluminum pedestrian bridge, called the Yards/Teague Connector Bridge, was constructed to provide access across the former barge slip and discharge outlet. At the western end of this bridge is a small landscaped plaza, squeezed between the O Street Pumping Station and the ballpark, called Diamond Teague Park. Similar to The Yards Park, the plaza is operated by Capitol Riverfront BID. At less than an acre in size, this small plaza provides public piers for canoes and kayaks.

FREDERICK DOUGLASS BRIDGE

The arrival of twenty-first-century bridge design became known to Southeast D.C. and Anacostia residents slowly in the late 2010s, as the gleaming-white curved arches of the new Frederick Douglass Bridge began to rise above the tops of the surrounding buildings.

The original Frederick Douglass Bridge at this location was another unattractive, utilitarian, midcentury highway bridge designed to move high volumes of commuter traffic into downtown. A steel-plate girder bridge constructed in 1950, it mostly served as an exit ramp for commuters from the Anacostia Freeway onto South Capitol Street, leading to Capitol Hill. The bridge had sidewalks, but they were narrow, and nonresidents rarely ventured on foot to the southern side of the river. In addition, structural weakness meant limiting truck traffic to the inside lanes of the bridge.

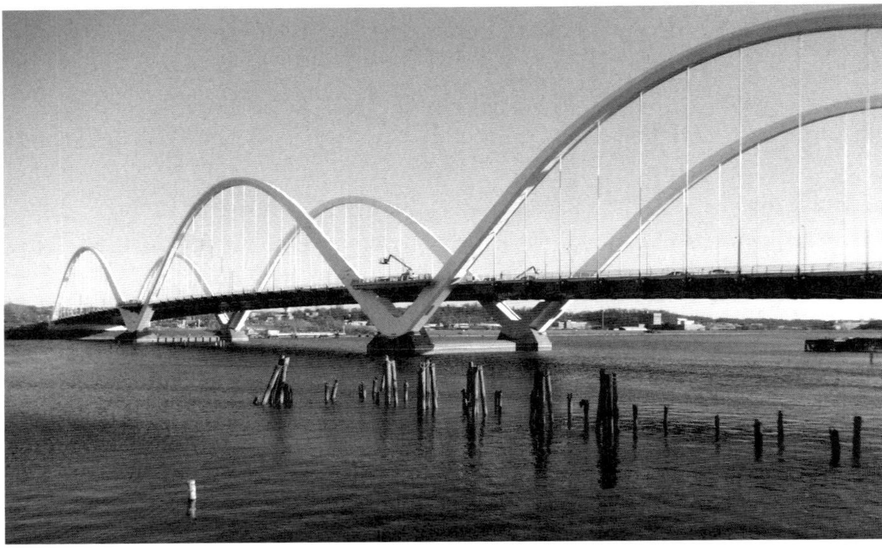

Final phases of construction on the Frederick Douglass Bridge in 2023.

In the early 2000s, the District, adjacent counties in Maryland, the National Park Service, the Environmental Protection Agency, and other government agencies embarked on a plan to clean and redevelop the entire Anacostia River and waterfront from Bladensburg to the Potomac. Perhaps the plan's most aspirational component was to lure a Major League Baseball team to a city that had already lost two of them—and then spend hundreds of millions to build a baseball stadium for a losing team at the end of an unattractive, industrial-looking bridge. All of this in a depressed, formerly industrial part of town that few suburbanites would willingly visit. On paper, the plan looked tenuous at best. The situation did not improve when the real estate market tanked into the Great Recession of 2008, the promised residential and commercial developments did not materialize, and the baseball team lost more than one hundred games for two years in a row.

By the mid-2010s, proving that urban renewal is a marathon and not a sprint, the Capitol Riverfront had become one of the most desirable neighborhoods in the city. And once this success had been established, the old, unattractive Frederick Douglass Bridge became a drag on further development. The new glass-sided high-rise condos and offices not only provided spectacular views of the ballpark and the river but also highlighted the bridge's aesthetic limitations. Meanwhile, few improvements were made south of the river, limiting the use of the bridge for pedestrian access to baseball games. In addition to its structural limitations, the bridge's aesthetic shortcomings were partially responsible for holding back development in the hottest real estate market in town.

In 2012, DDOT began planning for the replacement of the bridge, a replacement that would fully embrace all of the precepts of twenty-first-century bridge design. It is not just aesthetically pleasing—its three high white arches serve as public art, deserving of its prominent location just down the street from the U.S. Capitol. And it not only serves as the downstream connection between the East and West Bank Trails but also has facilitated the spread of redevelopment to the southern side of the river by providing easy and safe pedestrian access between the neighborhoods south of the river and the developments of the Capitol Riverfront and the ballpark. At each pier, between the arches, the sidewalk widens out into an observation platform, providing a view of new developments along The Yards Park.

The modernistic arches were a shock for many longtime District residents. For more than one hundred years, nearly all construction in the District had been low to the ground to avoid any disruption to the vistas of the

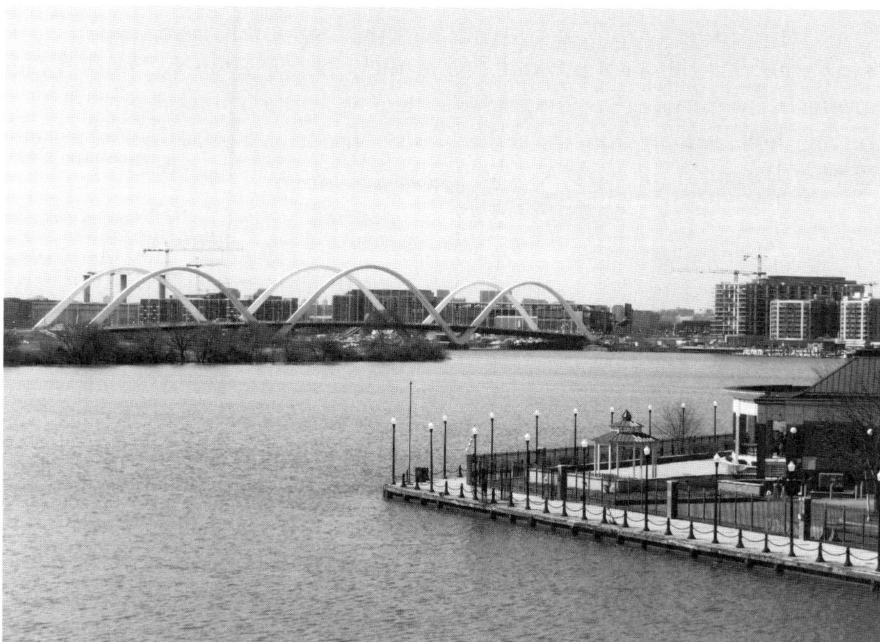

View of the Frederick Douglass Bridge, with the Navy Yard in the foreground, taken from the observation platform on the Eleventh Street Bridge.

Capitol Dome and the Washington Monument. In addition, most of the District's well-known, large-scale bridges were either classical in style, like Key, Arlington, and Taft Bridges, or midcentury eyesores like the Fourteenth Street and Eleventh Street complexes. But here, with little fanfare, rose a gleaming, white, modernistic sculpture, unlike anything else in the city.

Even more attractive, and again slightly shocking to those who think of the District's architecture as nothing but white marble and granite, is the bridge's lighting at night. The use of color-changing LED lights on bridges is now a common practice in many cities. The arches of the Frederick Douglass Bridge are not only modernistic in style and prominently visible, but they are also brightly lit in pastel shades of purple, orange, pink, and red. This is a significant departure from the bright white lights on the monuments, creating a distinct contrast between the traditional appearance of the city along the Potomac and the thriving, modern appearance of the city along the Anacostia.

A pleasing feature of the arches is the smooth continuation of the V-shaped concrete piers. The bridge's profile does not have any sharp angles. Instead, the superstructure is almost serpentine, with each arch descending smoothly to a curve sitting on top of the pier and then smoothly rising again to the next arch. The effect is pronounced at night because the arches and piers are lit in bright pastel colors, but the deck is not. This makes the deck all but invisible, leaving only the series of smooth, bright, pastel-colored arches displayed against the dark-sky background.

6

ROCK CREEK PARK AND NATIONAL ZOO

As discussed in chapter 1, the District encompasses dozens of small, miscellaneous parks that are owned and operated by the National Park Service. The NPS often consolidates small parks under its administration into larger units to make them easier to manage. Nowhere is this more true, and more potentially confusing, than the agglomeration of small and large parks called Rock Creek Park. In addition to the largest area that locals know as Rock Creek Park, approximately one hundred small, individual historical sites—former military forts, tributaries to Rock Creek, neighborhood parks, and traffic-jammed parkways throughout the District—are also incorporated within the larger administrative entity known as Rock Creek Park.

Because the various individual parks within the larger administrative unit are so widely distributed and have different geographic characteristics, the bridges are of different types and sizes and cannot be easily visited as a group. For the purposes of this guide, the bridges in Rock Creek Park are described in three separate chapters. Chapter 6 addresses the earliest and largest subunit within the larger administrative area known as Rock Creek Park. This northern section of the park is the area that most Washingtonians mean when they refer to "Rock Creek Park." In addition to the park itself, chapter 6 includes the portion of Rock Creek within the boundaries of the National Zoo. Although they are adjacent to each other, Rock Creek Park is administered by the National Park Service, while the zoo is part of the Smithsonian Institution.

IMPROVED FORDS AND EARLY BRIDGES

Because the Potomac and Anacostia are deep and wide estuaries, early settlers needed to cross these barriers on boats or ferries. However, Rock Creek was only a few inches deep and about fifty feet wide in most places, and the creek's rocky bottom offered a solid footing. Crossing Rock Creek by boat was not an option for early settlers, but the creek could be forded in many locations. These fords were often improved by paving directly across the bottom of the creek. It was even possible to construct small-scale bridges across the creek at multiple locations to support commercial enterprises.

Like the bridges on the Potomac and Anacostia, the history of the bridges on Rock Creek was one of repeated cycles of construction, destruction, and reconstruction. Multiple generations of bridges have existed at most of the current bridge locations, but a subtle difference exists between Rock Creek and the larger rivers. On the larger rivers, the replacement of a bridge was always a large-scale operation, usually requiring complete removal of all remnants of the earlier bridge. The piers needed to be removed because the river was used for shipping. The abutments needed to be removed to make way for roads and highways along the shores. The only exception is the lone remaining arch and pier of the Potomac Aqueduct in Georgetown.

Rock Creek Park was a different story. Reconstruction of new bridges was a small-scale operation and did not necessarily mean that older remnants needed to be removed. In most cases, the deck of a new bridge could be constructed on the piers of the older bridge, and in any case, there was no need to remove piers or abutments of former bridges to clear the way for shipping or new shoreline-based developments. On the contrary, the National Park Service has, from its early days, considered historic preservation one of its primary missions. Where a bridge abutment on the shore of the Anacostia may need to be removed because it is in the way or a visual eyesore, an abutment on the shore of Rock Creek may be preserved for its historic value.

As a result, Rock Creek Park still has multiple remnants of its former fords and bridges. None of these remnants are major attractions, and most will be unnoticeable to park visitors. However, many of these remnants are visible and can be investigated by visitors interested in the District's commercial, industrial, and transportation history.

The historic fords range from locations known only from historical sources to improved fords that could still be used by vehicles today. As an example of the former, the operations at Peirce Mill were supported by a ford until a bridge was constructed in the 1860s. The 1860 bridge was then replaced by later generations of bridges, and there is no obvious trace of the ford today.

Besides Peirce Mill, fords were improved at Blagden Mill, Milkhouse, and Klingle Road. The ford at Blagden Mill, also known as Argyle Mill, was in operation by the 1850s. The ruins of the ford, mill, and later bridges are located along Beach Drive between Picnic Grove 3 and Boulder Bridge. The mill and bridge were destroyed by the flood of 1889, leaving only an abutment of the bridge on the eastern bank of the creek. There is a historic information placard, but it is weathered and mostly illegible.

The current Blagden Avenue Bridge was the location of the famous Pebble Dash Bridge, celebrated as one of the most decorative small bridges in the park. Constructed in 1902, this was a small concrete arch bridge that carried Beach Drive across Broad Branch before the road turned east and crossed Rock Creek on the Old Rustic Foot Bridge. The Old Rustic Foot Bridge was constructed in 1907 and removed in 1939. It appears that it was not reconstructed immediately, because Zack Spratt's "Rock Creek's Bridges," published in the *Records of the Columbia Historical Society* in 1953, stated a footbridge was needed at this location because the ford was insufficient. Ultimately, both the Pebble Dash Bridge and Old Rustic Footbridge were replaced by the current bridges in 1957.

Milkhouse Ford is located about three-quarters of a mile north of the Military Road crossing on Rock Creek. It was used as a natural crossing and then improved by paving with six- to eight-inch-thick concrete in 1904. The roadway approaches on each side of the creek were also paved, originally with granite blocks and later with cobblestones, which are still in place today.

On the western side of the creek, adjacent to the paved approach, are the remnants of a rustic granite structure that holds a bench, apparently to allow people crossing the stream to rest and enjoy the scenery before continuing on their way. The ford was bypassed by a bridge in 1926, and the current Milkhouse Ford Bridge, which carries Beach Drive across the creek, was constructed in 1957. However, the ford was still open to the public in the mid-1990s. The ford is north of the historic 1883 Miller Cabin, which was moved to this location in 1912. It is one of the park's major attractions.

Documentation on Klingle Ford is not readily available, although the remains of it are easily visited today. Prior to the construction of the Rustic Log Bridge in 1895 and the Harvard Street Bridge in 1901, vehicles traveling from south of the zoo north had to cross a ford onto the zoo's property, continue on the zoo's road network on the western side of the creek, and then recross another ford north of the zoo. The Lower Ford was located on

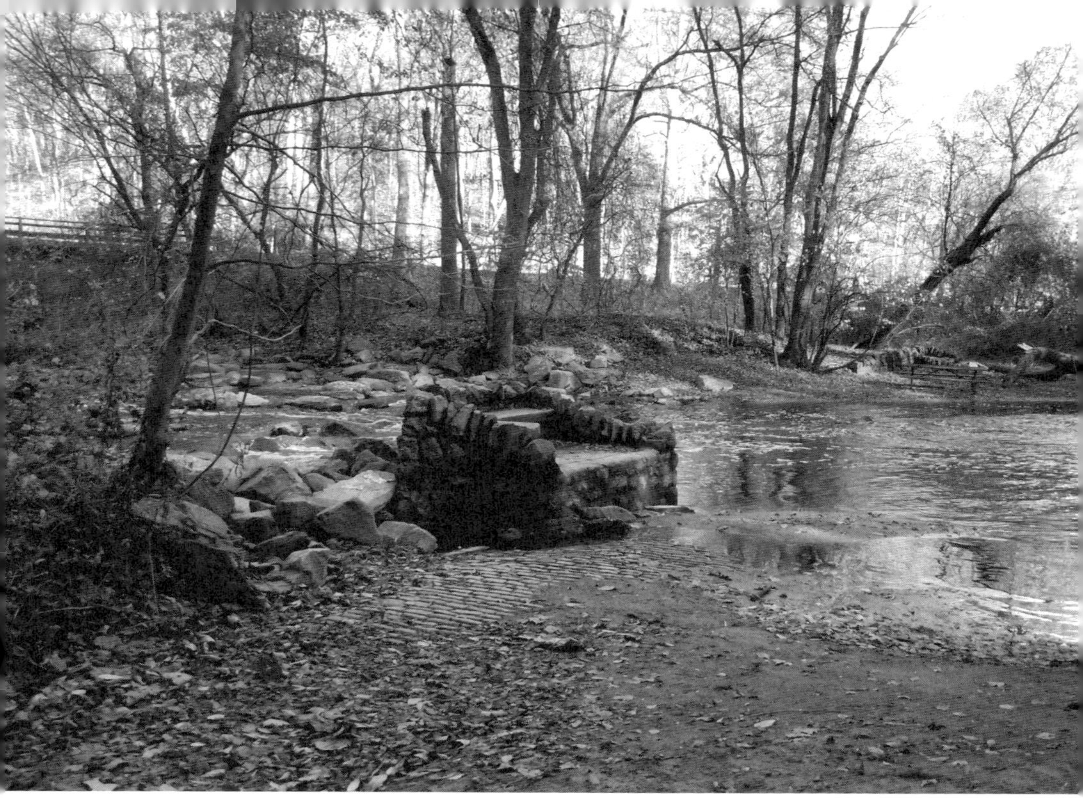

Remnants of cobble roadway and bench at Milkhouse Ford.

the southern edge of the zoo at the Adams Mill, named after President John Quincy Adams, who had owned the mill as early as 1825. These fords were also a popular bathing spot for nearby residents.

The Upper Ford was reported to be paved with concrete and may be the same as the remains of a ford found at the Klingle Road Bridge at the base of the Klingle Valley. The remains consist of rustic stone wing walls leading into the creek from its western side. There is no obvious trace of the improvements within the creek or on the eastern side of the creek.

In writing about the park's bridges in 1953, Spratt described an iron Howe truss footbridge crossing Rock Creek near the Jusserand Memorial between Peirce Mill and Bluffs Footbridge. He explained that it had been constructed at the Land Office Building on Eighth Street sometime between 1839 and 1860 and had been moved to this location south of Peirce Mill in 1901. He wrote about it in the present tense, suggesting it was still in place in 1953, although it no longer exists. A 2023 National Park Service inventory of its historic bridges in Rock Creek Park lists "Jusserand Foot Bridge Remnants," but there are no obvious ruins to be found.

HIKING TRAIL BRIDGES

There are many pedestrian-only trail bridges scattered throughout the park, and most are utilitarian and unremarkable. However, there is a set of historically important bridges, both in the Rock Creek Park segment and along the Rock Creek and Potomac Parkway segment, that deserves attention. These footbridges were funded by the Public Works Administration (PWA) and designed by the NPS during the Depression.

The PWA was a New Deal–era program established in 1933 to provide jobs by funding the construction of public buildings and infrastructure. The PWA funded many major projects throughout the District. In Rock Creek Park, the PWA was responsible for the restoration of Peirce Mill, the construction of Piney Branch Parkway, and other general improvements to the road network.

The PWA also left its mark on the network of footbridges used to connect the hiking trail system. In August 1933, a major storm flooded Rock Creek Park with two to four feet of water, damaging multiple footbridges beyond repair. This occurred only two months after the PWA had been established, and its staff had begun to look for projects. The NPS and PWA got to work immediately. The footbridges were designed by the Branch of Plans and Designs of the NPS's Landscape Division, according to the newly developed concept of "parkitecture." Although the bridges were constructed of durable, inexpensive concrete, the concrete arches sit on rustic-looking stone piers and are shaped and colored to fit into the surrounding landscape. Also, despite minor differences in specific details, they are all similar in appearance, tying them into a unifying aesthetic theme.

Because 1934 was not that long ago, it would seem easy to present a list of which bridges were constructed, but the documentation is not entirely clear. In 1939, C.W. Short and R. Stanley-Brown of the PWA wrote a summary of the PWA's public buildings and architecture from 1933 to 1939. They refer to five footbridges in Rock Creek Park as part of the PWA program and include photos of two of them: Riley Spring and Rapids Footbridges.

Spratt refers specifically to the following six footbridges funded by the PWA: Western Avenue Bridle Path (probably the same as the current Boundary Bridge), Riley Spring, Rolling Meadow, Rapids, Bluffs, and Saddle Club (also known as Shoreham Hill Footbridge). He also provides a 1934 construction date for all six. This is corroborated, mostly, by dates stamped into the concrete sides of three of the bridges: Riley Spring, Rapids, and Bluffs. Although Spratt provides a 1934 date for Boundary Bridge, the date stamped on its side is 1935. The Devil's Chair Footbridge, also known as Lyon's Mill Footbridge, is a curious case. Spratt mentions it, and it is of the same concrete arch construction type as the 1934 bridges, but Spratt gives a 1932 construction date and makes no mention of the PWA. Because 1932 predates the PWA and the flood, it seems likely that the 1932 date in Spratt's report is a typographical or other error.

In 1990, William Bushong developed a Historic Resources Study of Rock Creek Park for the Park Service, focusing only on structures north of the zoo. This document provided descriptions, and a summary of the historic significance, of five of the bridges: Rapids, Rolling Meadow, Riley Spring, Boundary (no mention of Western Avenue), and Bluffs. It left out Saddle Club and Devil's Chair, presumably because these are south of the zoo. This would still only total seven footbridges constructed using PWA funding in 1934. However, the descriptions of Rolling Meadow, Riley Spring, Boundary, and Bluffs all specifically state that there was a series of eight footbridges constructed using PWA funding in Rock Creek Park.

The most recent sources are the National Park Service's own National Historic Preservation Act documents dating from the 1990s. There are seven individual Historic American Engineering Record (HAER) reports for the five bridges north of the zoo (Bluffs, Boundary, Rapids, Riley Spring, and Rolling Meadow) and the two bridges south of the zoo (Devil's Chair and Saddle Club). All seven of these individual reports state that there were eight footbridges funded by the PWA, but none of them lists the names of eight bridges.

A possible explanation for the missing eighth bridge could be the concrete arch footbridge at Peirce Mill, which is of the same construction type and appearance as the 1934 bridges. However, the Peirce Mill footbridge was constructed in 1982, as documented by the date stamped into the concrete sides. It was clearly designed to match the PWA bridges as part of a unifying theme, but was constructed almost fifty years after the earlier bridges.

Although each of the eight footbridges is unique in its details, they are all of the same basic construction type and appearance. Each bridge is a low concrete arch, about one hundred feet long, and placed on stone abutments, wing walls, and piers to present a rustic appearance to park visitors and hikers. The concrete deck forming the arch is of a uniform thickness on some of the bridges, flat on both the top and bottom. On other bridges, the top is flat, but the bottom is scalloped, with the deck being thicker at the abutments and piers and thinner in between. All but one of them have a post-and-rail railing, evoking the appearance of a rustic split-rail fence, but it is only after close examination that the railings on some of the bridges can be seen to be made of concrete and only treated to look like wood.

Their most striking similarity is that they all cross a lovely babbling stream flowing through the woods and rocks. From upstream to downstream, the footbridges are Boundary, Riley Spring, Rolling Meadow, Rapids, Peirce Mill, and Bluffs within Rock Creek Park, and then Saddle Club and Devil's Chair adjacent to Rock Creek and Potomac Parkway.

Boundary Bridge is exactly as it is named, crossing Rock Creek at the Maryland-District boundary. Because the northwestern Maryland-District boundary is marked by Western Avenue for most of its length, this is likely the bridge referred to by Spratt as the Western Avenue Bridle Path Bridge, even though Western Avenue dead-ends before it reaches this location. The Valley Trail crosses the creek on Boundary Bridge a few hundred feet east of a parking lot off of Beach Drive.

Riley Spring Footbridge, located about a mile south of Boundary Bridge, carries a connecting spur trail from the Valley Trail across the creek to connect to the Western Ridge Trail on the opposite side of Beach Drive. It is a symmetrical bridge on two stone piers, consisting of three concrete arch spans that are flat on top but scalloped on the bottom. The railing appears to be a standard wooden post-and-rail but is made of concrete, with decorative shaping on the tops of the posts over the abutments and piers.

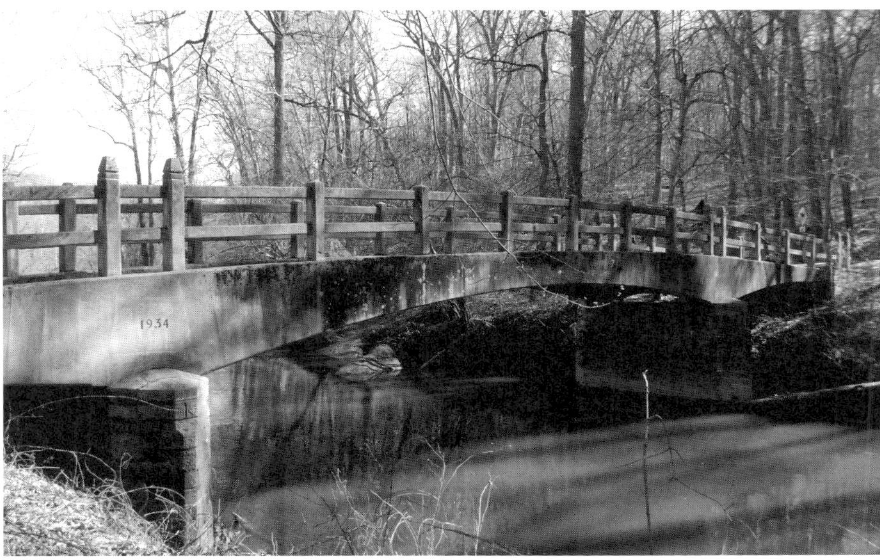

Riley Spring Bridge, displaying construction date of 1934.

Rolling Meadow Footbridge carries a spur trail from the Valley Trail east of the creek to Beach Drive on the west. Its construction type is a bit unusual. It is a single-span concrete arch but rests entirely on piers instead of abutments. The bridge is the only one of the group that does not have a post-and-rail railing. Instead, it has a steel railing, with the end-posts shaped into a curved art deco design. It also does not have a date stamp.

The Rapids Footbridge, located a mile north of Peirce Mill, carries a spur trail from the Valley Trail and Beach Drive east of the creek to another trail west of the creek. If there can be one bridge in a group of largely similar bridges that is more special than the others, Rapids Footbridge is that bridge. The distinction is subtle, but Rapids has a few decorative flourishes that make it slightly more appealing than the other PWA footbridges. Even the 1992 HAER reports for the other footbridges refer to Rapids Footbridge as being the "most notable" of the group. The concrete deck is three spans long but has a uniform thickness from bank to bank. It rests on two stone piers and two abutments, but the piers are different from those on the other footbridges. Instead of the concrete deck resting directly on stone piers, the piers are topped with a concrete capital. The concrete capital spreads the load of the deck across a wider area than just the top of the pier and is also decorative, with a geometric art deco design. The railing is wooden post-and-rail, but with decorative, carved finials on top of the posts above the abutments and piers. The 1934 construction date is stamped on both sides. In addition to the decorative flourishes, this may be one of the loveliest settings among the footbridges, with a view of nothing except wooded hillsides and the creek flowing through a field of boulders. The bridge is on a section of Beach Drive between Broad Branch Road and Joyce Road that is closed to traffic and is only accessible on foot or bicycle. The closest parking is on Joyce Road at Picnic Area 22, requiring a one-way hike or bike ride of about a mile. Despite this substantial distance, you will still find it difficult to get good photos of the bridge on fall foliage weekends because of the enormous crowds that go to the bridge to enjoy the scenery.

Bluffs Footbridge is located between Peirce Mill and Piney Branch Parkway just north of the zoo. The bridge's name derives from the valley's narrow width at this location between the steep hill slopes. The bridge is asymmetric, with a longer scalloped concrete span on the east, a shorter scalloped span on the west, and a single stone pier in the middle of the creek.

The Saddle Club and Devil's Chair Footbridges are located on the asphalt trail adjacent to the Rock Creek and Potomac Parkway. Saddle

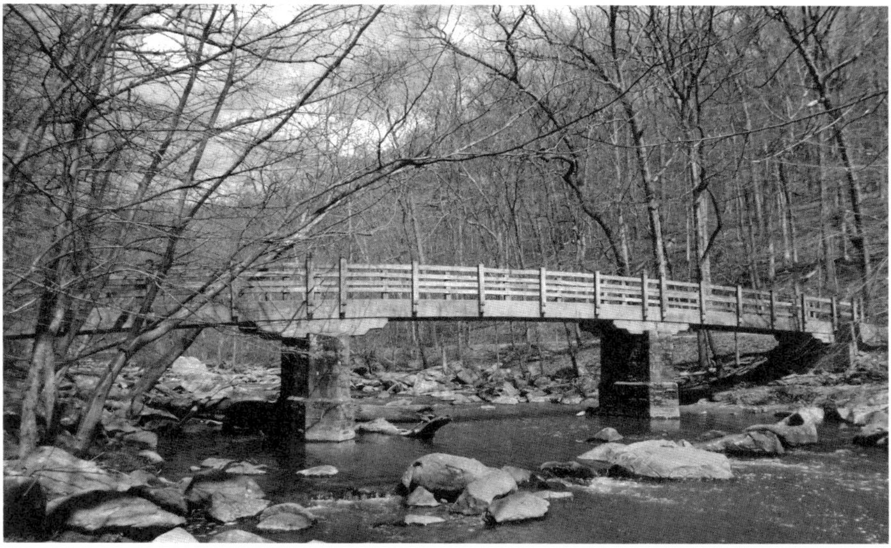

Rapids Bridge, considered the most scenic of the hiking bridges constructed in Rock Creek Park in 1934.

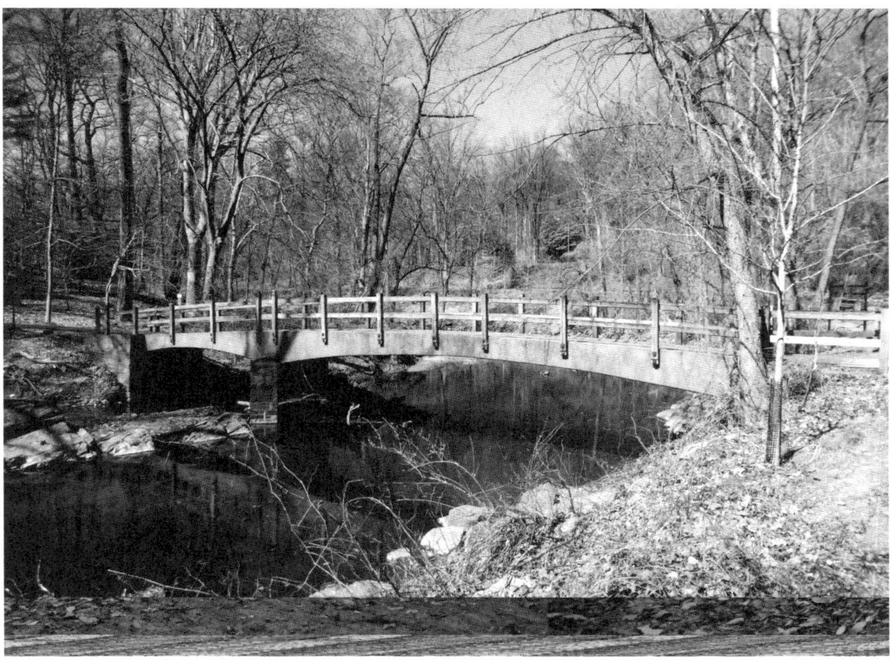

Bluffs Footbridge crossing Rock Creek a short distance south of Peirce Mill.

Club is located just south of a complicated series of on-ramps near the Taft Bridge that mark the transition from Beach Drive to parkway. Devil's Chair is located between the historic Oak Hill Cemetery on its southern end and Embassy Row along Massachusetts Avenue on the north, but neither of these are accessible from the bridge or trail. Although they support a continuous hiking and bike trail all the way to Georgetown, the Saddle Club and Devil's Chair bridges are not as rustic or appealing as the five bridges in Rock Creek Park proper. The Rock Creek and Potomac Parkway is a much busier highway than Beach Drive and is open to traffic at all times. Both bridges need to be accessed by hiking or cycling the trail, as neither of the bridges is situated near parking access.

BOULDER BRIDGE

Boulder Bridge, located a quarter of a mile south of Rapids Footbridge, is one of the most unusual and photogenic bridges in Rock Creek Park. It is also one of the most historic, having been constructed in 1902, only twelve years after Congress established the park. The bridge is one of only two bridges in the park, and seven bridges in the entire city, to be listed on the National Register of Historic Places.

Boulder Bridge is one of the earliest examples of parkitecture in the city. Structurally, it is an unusual Melan arch bridge constructed by encasing a steel arch bridge in concrete. The bridge is twenty-three feet wide, eighty feet long, and twelve feet high above the creek. It carries Beach Drive across Rock Creek, although Beach Drive is closed to traffic at this location.

A remarkable feature of the bridge, and the source of its name, is that the stones used for the facing of the spandrels and parapets are not flat. Instead, they are enormous rounded boulders. This design was proposed by Colonel Lansing H. Beach, the namesake of Beach Drive, who was the secretary of the Board of Control of Rock Creek Park at the time.

Large round boulders were used with the intention that they would help the bridge blend in with its surroundings by mimicking the shape and size of the boulders found in the creek. However, the boulders used were substantially larger than those on the ground or in the stream. The anecdotal explanation is that Colonel Beach directed that "man-sized" boulders be used, intending that "man-sized" meant that they could be carried and placed by a single stonemason. This would mean that each

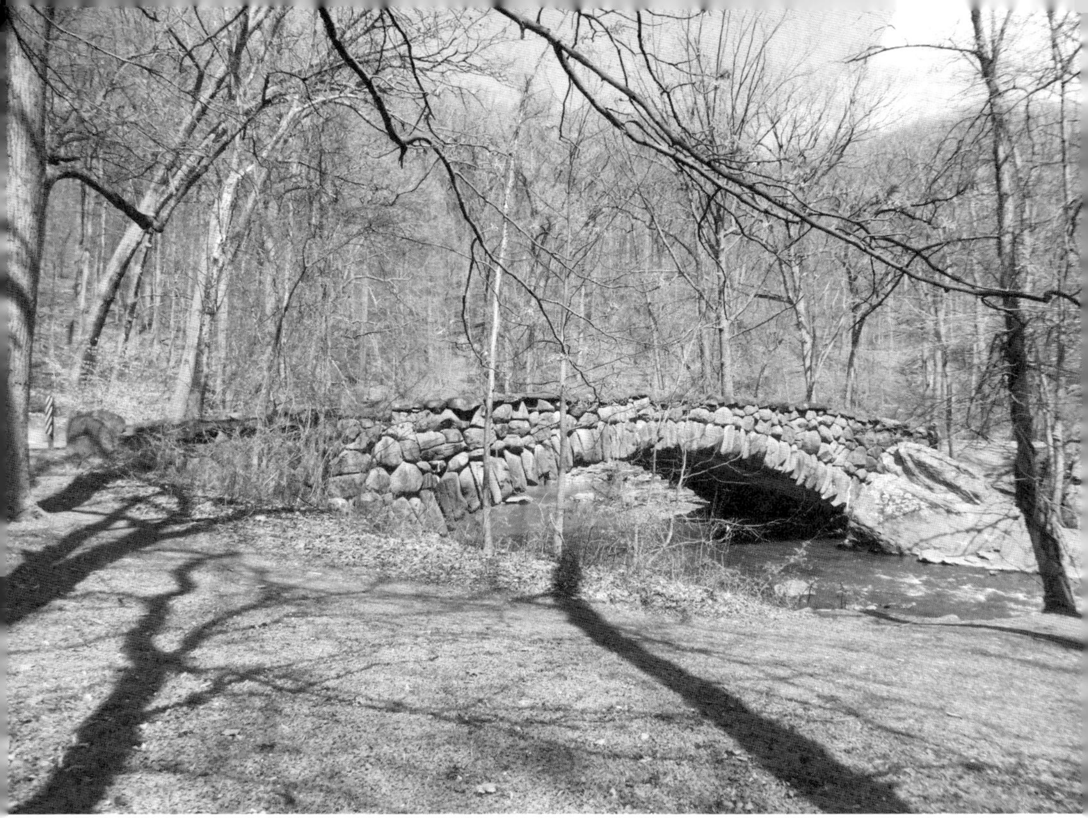

View of Boulder Bridge in spring.

stone weighed about twenty to thirty pounds. Instead, the contractor interpreted the contract as requiring stones as large as a man, which would weigh one hundred or more pounds each. By the time Colonel Beach visited the site and noticed the error, construction was too far along to start over, so the bridge was completed with the gigantic stones.

Photographs of the bridge do not give a proper impression of the size of the boulders. One subtle feature to notice when walking across are the dozens of small holes drilled into the stones, usually about two or three per stone. These holes are tiny, only about a quarter-inch wide and a half-inch deep, and they are barely noticeable unless you know how to look for them. However, they were clearly drilled and were likely needed to allow the quarry workers and masons to use tools to grasp, lift, and maneuver the boulders into place.

The picturesque bridge's scenic setting makes it one of the park's more popular destinations for rustic hiking, photography, and picnics. However, because Beach Drive is closed to traffic between Broad Branch Road and Joyce Road, it is not directly accessible by car. The closest parking lot is

Picnic Grove 2a North, near Peirce Mill, and accessing the bridge requires a one-way bike ride or walk of about one mile. Rapids Bridge and Boulder Bridge are probably the two most scenic hiking bridges in the entire park, and because they are only an easy ten-minute walk apart, they should be seen together.

PEIRCE MILL AREA

Peirce Mill is not necessarily the most important location for tourist bridges in Rock Creek Park. However, it is probably the park's most prominent attraction for history-based tourism and recreation, and the area does have several bridges that provide access to various sites.

Isaac Peirce purchased 160 acres and an existing mill on Rock Creek in 1794, about a mile north of what is now the zoo. The primary businesses at that time were apple and various other fruit orchards. In about 1800, Peirce established a sawmill, and the operation continued to expand. The mill building, which can be visited today, was constructed in 1829. The machinery inside the building was restored in 1935, and demonstrations are given on Saturdays. An 1801 springhouse, which is a small building constructed over a spring, is still in place, oddly sitting in the median between two lanes of Tilden Street. The carriage house adjacent to the mill dates from 1810. By the time Isaac Peirce died a wealthy man in 1841, the industrial complex in what is now Rock Creek Park amounted to more than 2,000 acres, including the mill, distillery, and farming operations.

Such an industrial operation on the banks of a substantial creek would obviously require a bridge to support transport of feedstock and milled products. A painting of the mill building, dated about 1830, shows a ford, which is reported to be one of the earliest crossings of Rock Creek. The earliest bridge, directly adjacent to the mill building, was constructed by the 1860s. It was replaced by a new wooden bridge named Shoemaker's Bridge in 1872. Although the bridge was originally constructed to serve the mill, the growth of the city and the establishment of Rock Creek Park in the 1890s led to increased use as an east–west thoroughfare through the park.

The bridge was reconstructed in 1895 by placing steel girders and a wooden deck on the original 1872 stone piers and abutments. Stone wing walls were added in 1912. In 1921, the eastern stone abutment was replaced with a new stone-faced concrete abutment, the deck was surfaced with

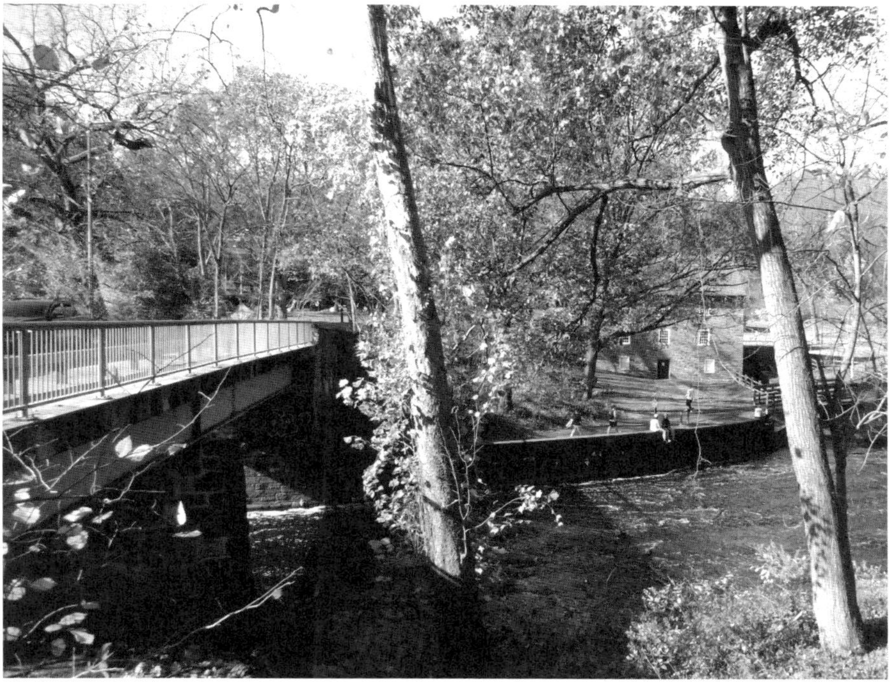

View of Peirce Mill and adjacent Tilden Street Bridge.

asphalt, and steel railings were added. However, the narrow width of the bridge posed a severe limitation on the increasing amount of traffic, so the current bridge is a 1993 steel-girder replacement with a pedestrian sidewalk.

In addition to being a major historical attraction, Peirce Mill is one of the most important access points for hiking and biking in the park because it can be easily reached by car. It is just across the creek from Beach Drive, and there are multiple roadway entrances to Beach Drive from the west at Broad Branch Road and Porter Street and from the east at Piney Branch Parkway and Klingle Road. The mill is also easily accessible from the major thoroughfare of Connecticut Avenue by using Tilden Street NW about a half mile to the mill. The Western Ridge Trail passes directly through the narrow space between the mill and the creek and then under the road bridge. At the southern edge of the picnic area, the trail crosses another concrete arch footbridge, designed to match the unifying theme of the 1934 PWA footbridges. However, this one was constructed in 1982. A short distance south of this footbridge, visible from the trail across Beach Drive, is the Jusserand Memorial Bench.

NATIONAL ZOO BRIDGES

Like the Potomac and Anacostia, Rock Creek flows across two distinct types of landscapes on its way to reaching sea level. The northern section flows across crystalline bedrock of the Piedmont, which is characterized by rocky stream beds and deep, steep-sided valleys. Once the stream crosses the Fall Line, it enters terrain that is relatively flat and underlain by soft Coastal Plain sediments. This transition can easily be seen on topographic maps of the city. Those maps show that the steep, rocky terrain of the Piedmont is marked by closely spaced topographic contours and mostly irregular road patterns, while the Coastal Plain has more widely spaced topographic contours and adheres to a more rectangular road pattern.

Along the route of Rock Creek, this transition occurs just south of a promontory known as Woodley Park. Both upstream and downstream of this location, Rock Creek flows almost directly north to south. However, at Woodley Park, the promontory forces the route of the creek east, then west, then north, and then south again, all within a distance of less than about three-quarters of a mile. It was on the eastern edge of this promontory where, in 1889, Congress chose to establish the Smithsonian National Zoological Park to house the animals that had, until then, been living in pens adjacent to the Smithsonian Castle on the Mall. Because of the steep slopes north and south of the Woodley Park promontory, a complex network of roads, trails, and bridges converges to cross the creek both north and south of the zoo.

In general, none of these bridges are particularly historic, decorative, or interesting. North of the zoo, the complex includes a network of interchanges between Porter Street, Klingle Road, and Beach Drive. These are attractive bridges with sidewalks constructed in the late 1940s, painted dark green, and resting on stone-faced piers, but they are of limited interest.

Directly adjacent and parallel to the Klingle Road Bridge is an unusual footbridge at the junction of the Rock Creek and Klingle Valley trails west of the creek. The two trails join, cross this footbridge, and continue south as the Rock Creek Trail in the narrow space between Rock Creek and Beach Drive. Among the Rock Creek bridges, this bridge's superstructure is unusual. The structure is a laminated wooden arch that extends about ten feet above the wooden deck. The bridge would be picturesque and an attraction for rustic selfies in the park except for its location underneath the less picturesque, dark-green steel girders of the Klingle Road Bridge. The bridge was likely constructed after 1991, as it is not listed in the National Park Service inventory of bridges in the 1991 National Register of Historic Places Registration Form for the Rock Creek Park Historic District.

South of the zoo, where the creek exits the Piedmont, the roadways and trails are even more complicated. For most of the twentieth century, the narrow width of the valley and sharp curve of the creek in this area forced traffic on Beach Drive to ford the creek into the zoo, traverse roads within the zoo, and then ford the creek again to reach the Rock Creek and Potomac Parkway south of the zoo. Zoo officials objected to this traffic as a disruption to their operations, as well as a disturbance to the animals, and opposed any proposals to improve the roadway within the zoo's property. Eventually, the solution was to straighten the road by boring a tunnel through the promontory that is the site of the historic Mt. Pleasant Plains Cemetery at Walter Pierce Park.

The tunnel, completed in 1966, bypassed the zoo, eliminating any need for traffic to enter the zoo's property. Immediately on exiting the tunnel, another relatively uninteresting midcentury road bridge carries Beach Drive across the creek again. The tunnel even includes a sidewalk, providing a connection between the trails north and south of the tunnel.

Even though the sidewalk through the tunnel is still accessible, it is not a particularly pleasant or safe walk or bike ride with the speeding traffic in a confined space. In 2021, the trail south of the zoo was improved by the addition of a modern pedestrian-only footbridge across the creek adjacent to the road bridge. The footbridge allows trail users to bypass the tunnel by following the trail around the promontory along the zoo fence line between Beach Drive and Rock Creek.

While the bridges crossing Rock Creek north and south of the zoo are not decorative or historic, the half-mile-long segment of Rock Creek within the zoo is another story. There are three unusual bridges within the zoo; two are more than one hundred years old and are among the most decorative small bridges in the District.

Despite being situated within one of the city's most popular tourist and recreational attractions, these bridges are not well-known and are rarely seen by visitors. The zoo's property is bounded on the west by Connecticut Avenue and on the east by Rock Creek. Prior to the 1960s, there were public vehicle entrances on both sides of the zoo, as well as an entrance for service vehicles on the eastern side. The vehicle entrances on the eastern side from Beach Drive needed to cross Rock Creek on short, low-level bridges to access parking lots on the zoo's eastern side.

The original public entrance on the eastern side of the zoo was across a famous bridge named "Old Rustic Log." This was a wooden arch bridge constructed in 1895, shortly after the zoo was established. The bridge was decorative and romantic in appearance and featured on postcards at the time. It was intended to be a temporary entrance, and a more solid bridge, an extension of Harvard Street, became the main entrance when it was constructed a few hundred feet to the north in 1901. The Old Rustic Log Bridge was replaced in 1913 by a single-span concrete arch bridge, which is decoratively faced with blue-gray gneiss stonework. Today, it is a service and staff entrance road for the zoo and is not technically open to vehicles or pedestrians. However, the Rock Creek Trail crosses the southern end of it, and the Amazonia attraction within the zoo is directly on the northern end, so the attractive bridge can be visited and photographed on foot. For some reason, the bridge is listed on Wikipedia as the "Jewett Street" Bridge, although this name does not appear in other sources.

A second bridge, constructed in 1901 as a vehicle entrance from Harvard Street, is now known as the "Old" Harvard Street Bridge. It is a single-span Melan concrete arch. Cream-colored, with decorative geometric shaping on the spandrel and a neoclassical balustrade, this bridge was clearly intended to serve as a grand, decorative entrance for visitors to the zoo.

Together, these two beautiful historic bridges are a reminder of the City Beautiful Movement, when aesthetics and decoration were used to improve the visitor experience for both Rock Creek Park and the zoo. Unfortunately, both bridges are now almost entirely hidden from view for not only most District residents, but also most zoo visitors. This is because traffic management considerations on Beach Drive and at the zoo have resulted in closing direct access to vehicles and pedestrians from Rock Creek Park into the zoo. Vehicle access is now limited entirely to the Connecticut Avenue entrance, almost a mile west of Rock Creek.

To see the bridges, pedestrians within the zoo can walk down to the far eastern boundary to the creek, but it is a long walk, and they can only do this if they know that there is something interesting to see. There are no historic information placards on either bridge. Both bridges are located near parking lots D and E, so a dedicated bridgespotter can visit by car. However, since vehicles entering at Connecticut Avenue must drive past lots A, B, and C to reach lots D and E, it is likely most visitors do not get to this relatively remote eastern corner of the zoo.

There is still a pedestrian entrance into the zoo across Rock Creek from the east, but it does not lead directly into Rock Creek Park. Instead, it crosses

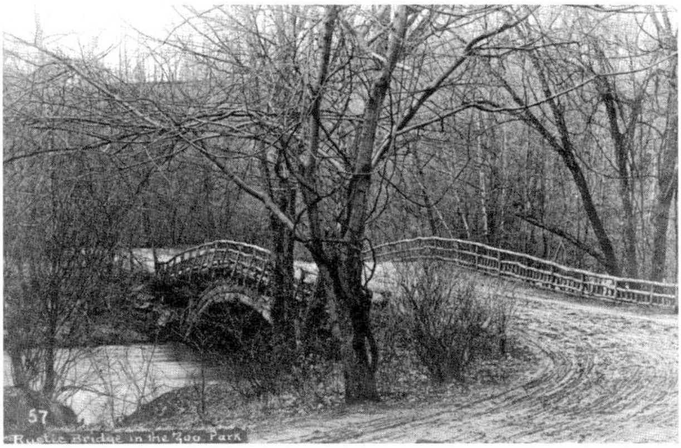

Left: View of the original Zoo entrance road bridge, known as the Rustic Log Bridge. *From collection of John DeFerrari.*

Below: The Zoo Service Road Bridge, also known as Jewett Street Bridge, leading into the National Zoo on the opposite side of Rock Creek.

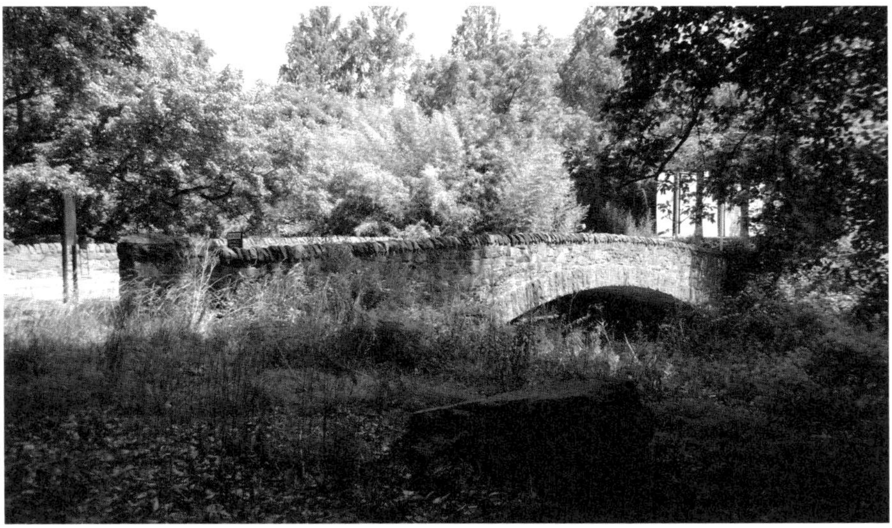

over both the creek and Beach Drive and connects to Harvard Street, making this the "New" Harvard Street Bridge. It is located only a few hundred feet north of the Old Harvard Street Bridge and is a good place from which to capture photos of the older bridge. The bridge was constructed in 1965 and is a rare trapezoidal concrete box girder, which would have been considered a futuristic-type design at the time. The bridge was constructed as a road bridge but is now closed to public vehicles. However, it is open for pedestrian access into the zoo from the Mt. Pleasant and Lanier Heights neighborhoods east of Beach Drive.

Although there are no information placards at these bridges, there is one, close to the Great Cats exhibit, which discusses the long-standing tradition of African American families visiting the zoo on Easter Mondays. This placard includes a historic photo of crowds crossing the gorgeous Old Harvard Street Bridge on Easter Monday in 1910. The photo is a reminder of the old days, when the zoo and the park were one attraction that could be visited together on a sunny afternoon. While the need to manage both zoo and Beach Drive traffic is understandable, it seems something was lost by closing off the direct connections between the zoo and the park.

7

ROCK CREEK AND POTOMAC PARKWAY

While driving, walking, or cycling through Rock Creek Park north of the zoo, you will pass through a quiet, rustic landscape of forested hillsides, a babbling creek, just a few visible structures, trails packed with runners and cyclists, and a few scattered low-profile road and footbridges. But on emerging into daylight at the south portal of the Zoo Tunnel, the entire world has changed.

At that location, where the trail through the tunnel merges with the trail along the creek adjacent to fence surrounding the zoo, the scenery instantly transforms. Instead of a rustic park, you are now firmly in the middle of the city. The valley is narrower, and buildings are visible along the tops of the bluffs. Just south of the Taft Bridge, quiet, casual Beach Drive merges with Cathedral Avenue and Shoreham Drive to become loud, frantic Rock Creek and Potomac Parkway. The speed of the traffic increases, even if the speed limit has not. The vehicles driven by casual park visitors looking at scenery are replaced by commuters who are anxious to get to the office.

And instead of boring, low-profile midcentury road bridges, you are immediately within the shadow of one of the highest, most historic, and most ornate bridges in the District: the Duke Ellington Bridge. And this is not a one-off. From this location and continuing south for two miles along the Rock Creek and Potomac Parkway to Georgetown, there are six more high, historic, and decorative bridges crossing the entire valley, one after another. They are all decorative, both in terms of form when viewed in profile, and in terms of applied decorations along the sidewalks. However, be advised that they cannot be easily viewed in profile on a driving trip. The traffic moves fast, and there is literally no place to stop, as there are no shoulders or observation areas. The only way to take the time needed to view these bridges from the side is on foot or bicycle from the Rock Creek Trail.

DUKE ELLINGTON BRIDGE

The Duke Ellington Bridge, which carries Calvert Street across the wide, deep Rock Creek Valley just south of the Zoo Tunnel, is one of three large-scale bridges constructed to facilitate the development of residential suburbs in Northwest D.C. and the Maryland suburbs.

Throughout most of the nineteenth century, the steep slopes of the Rock Creek Valley served as a barrier to overland travel between the Northwest D.C. suburbs and downtown. In the late 1880s, a wealthy developer named Francis Newlands, later a senator from Nevada, formed the Chevy Chase Land Company and began purchasing land west of Rock Creek. To connect the land to the center of the city, he also formed the Rock Creek Railway, which constructed the Calvert Street Bridge across Rock Creek to carry streetcars west to Connecticut Avenue. After crossing at Calvert Street, the streetcar line turned northwest on Connecticut Avenue and followed this all the way to the new village of Chevy Chase in Maryland. Both the original Calvert Street Bridge, constructed in 1891, and the current bridge, constructed in 1935, were specifically designed as streetcar bridges to carry the Rock Creek Railway line into the northwestern suburbs and Maryland.

The original Calvert Street Bridge at this location was a six-span-long iron trestle bridge more than 120 feet high. That bridge was beginning to fail as early as 1911, leading to a more than twenty-year-long process of reinforcements while the design of a replacement bridge was debated.

It is difficult to discuss the design of the 1935 Ellington Bridge before that of the 1907 William Howard Taft Bridge that carries Connecticut Avenue, which is the next high-profile road bridge to the south. Because Calvert Street and Connecticut Avenue are oriented differently, the western ends of the two bridges almost meet at a single point at the intersection of Calvert and Connecticut, even though their eastern ends are separated by more than a quarter mile. This crossing at odd angles, as well as the extreme height of the two bridges, places both bridges within the same viewscape for drivers on the bridges, drivers under the bridges, and residents living in apartment buildings lining the top of the hillsides. During the period from 1907 to 1935 and even today, that viewscape was dominated by the ornate, open-spandrel concrete arches and sculptures of the enormous Taft Bridge.

This created a delicate balancing act for the design of the replacement of the Calvert Street Bridge. By 1917, bridge design in Washington, D.C., was directly subject to approval by the U.S. Commission of Fine Arts. The new bridge design had to be concordant with the decorative concrete arches of its larger neighbor, the Taft Bridge. However, it could not be larger, or more decorative, or in any other way overshadow the Taft Bridge, which was, at the time, an important District monument and the largest unreinforced concrete bridge in the world. The Calvert Street Bridge had to be special, but not too special.

This balancing act led to almost fifteen years of competition and debate over the new bridge, all while the old bridge continued to fall to pieces. In 1917, a local architect, George Oakley Totten, submitted sketches for a five-arched concrete bridge, consisting of three large arches flanked by two smaller arches, all with elaborate decorations. The U.S. Commission of

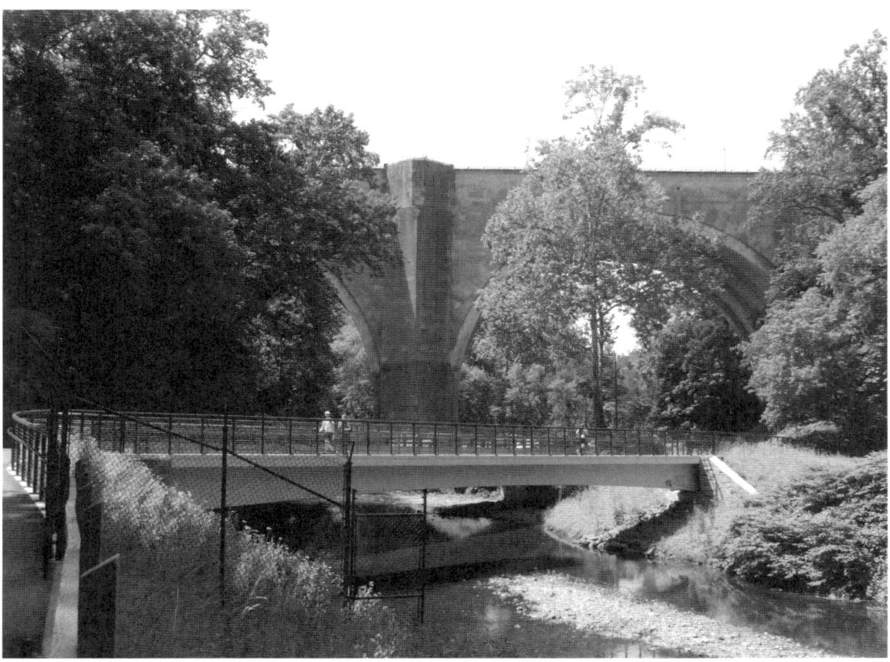

The pedestrian bridge south of the Zoo Tunnel, with the Duke Ellington Bridge hovering in the background.

Fine Arts gave tentative, lukewarm approval but made no efforts to move forward. In 1931, another architect, Paul Cret, proposed a single-span steel arch design with toned down ornamentation. Congress intervened, placing the decision under the District commissioners, who initially approved Cret's design but then determined that a steel bridge would generate too much traffic noise. In 1933, the U.S. Commission of Fine Arts once again stepped in and insisted on multiple concrete arches instead of steel. Totten reproposed his 1917 design, which was rejected, again, for being too ornate. With the funding already approved and construction ready to begin, the District commissioners worked with Cret to finalize a three-span concrete arch faced with Indiana limestone.

The construction of the bridge was unusual. To allow uninterrupted streetcar service during the construction, the existing steel trestle bridge was placed on rollers and moved, slowly by horses driving winches, about eighty feet to the side. Streetcar traffic continued over this bridge while the new bridge was constructed at the original location.

In keeping with the balancing act, the Ellington Bridge has sculptural decorations, but they are relatively muted compared to those on the nearby Taft Bridge. The decorations are bas-reliefs carved into the sandstone facings of the monuments on the four corners of the bridge, along the sidewalks. The bas-reliefs, each about three feet square in size, are allegorical depictions of gods and goddesses representing the four modes of transportation that symbolized modern progress when the bridge was constructed in 1935.

Travel by air is depicted on the southwestern corner, nearest the Taft Bridge, as a male nude in the clouds, held aloft with modern airplane wings. This monument also has a bronze plaque renaming the bridge for D.C. native Edward Kennedy "Duke" Ellington. Directly across Calvert Street, the carving on the northwestern monument shows a god with a trident, floating across stone-carved waves, with a ship in full sail behind. The carving on the southeastern monument depicts a woman with flowing hair while a speeding car with 1930s-type fenders kicks up a dust cloud. Finally, the carving on the northeast is a god depicted as a train engine, gliding along a stone-steel rail over stone-wooden ties and enveloped in steam. This northeastern monument also has a dedication plaque to Duke Ellington and sits at the entrance to Walter Pierce Park.

The Ellington Bridge is an extremely comfortable bridge to walk across. The bridge originally carried both traffic lanes and a streetcar line. After removal of the streetcar line, the three traffic lanes remained, and the

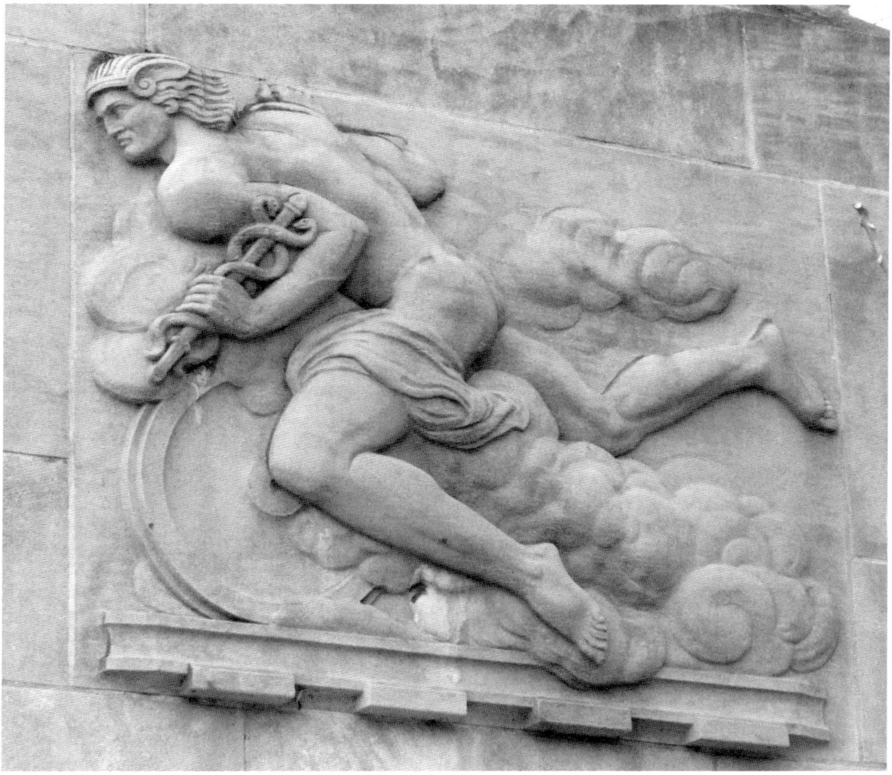

Close-up view of art-deco carving on one of the four corner monuments at the Duke Ellington Bridge.

remainder of the width of the bridge was dedicated to pedestrian use, making the deck of the bridge seem more of an urban plaza than a bridge with sidewalks. At the piers, the sidewalks open out to small observation areas with great views out over the Rock Creek Valley and especially magnificent views of the gorgeous open-spandrel arches of the Taft Bridge a short distance away to the south. To the north, the vista is effectively an aerial view of traffic exiting the south portal of the Zoo Tunnel and then immediately being funneled over one of the low-profile road bridges over Rock Creek. The view is so attractive that it is worth quoting the description written in 1953 by Spratt, which refers to the "unsullied vista of sylvan beauty," "pleasing panorama," and "lovely expanse of sylvan verdure."

As with the other high-profile bridges south of the zoo, there is no pedestrian access from the sidewalk of the bridge down to the trail below. The arch openings are enormous, towering more than sixty feet above the traffic and cyclists. The closed spandrels are beige in color and smooth, with no texturing. Even the interior of the arches is decorative, with the four ribs of the arch each protruding out about a foot from the otherwise smooth facing of the arch.

EDGEWATER STABLES/WOODLEY LANE BRIDGE

Almost unnoticeable, squeezed in between the closely spaced giants of the Ellington and Taft Bridges, is a low-profile bridge crossing Rock Creek, carrying a driveway from the parkway into the U.S. Park Police Training Center at Edgewater Stables. The driveway is marked by a sign reading "Authorized Vehicles Only," and the bridge would not appear to be of any historical or decorative interest. However, one of the features listed on the National Park Service inventory of its historic bridges in Rock Creek Park is "Old Woodley Lane Bridge Remains," which were once found at the end of the driveway.

The Woodley Lane Bridge was the predecessor to the current Taft Bridge. Constructed in 1889, it was called the "High Bridge" because, until the 125-foot-high Taft Bridge was constructed in 1907, its fifty-four-foot height was considered "high." The bridge was a steel truss, four spans long, and carried vehicles and sidewalks. The northern end of the bridge provided access to Woodley, the estate of President Grover Cleveland.

Although the 1992 Historic American Engineering Record Form for the Woodley Lane Bridge claims that the bridge carried streetcars, there is no statement to that effect in earlier source documents, including Emery in 1938 and Spratt in 1953. The 1897 map of the Rock Creek Railway (by then named the Capital Traction Company) shows that the streetcar crossed the creek on Calvert Street, turned northwest along what is now Connecticut Avenue, and then crossed Klingle Valley north of the zoo. The same thing is seen on the 1900 topographic map of the District, which clearly shows streetcar lines differently from roadways. On this map, there is no continuous connection of Connecticut Avenue across Rock Creek, and the Woodley Lane Bridge is shown carrying a roadway, but not a streetcar, across Rock Creek at a different angle than the current Taft Bridge. Therefore, the Woodley Lane Bridge never carried streetcars. The Taft Bridge, constructed in 1907, was the first creation of a straight, continuous roadway-only connection between downtown D.C. and the Maryland suburbs.

The bridge was demolished after the Taft Bridge was completed. Although there is no vehicle access into the Edgewater Stables, there is a crosswalk allowing pedestrians and cyclists to cross the parkway from the Rock Creek Trail. While no remains are to be seen, there is a historical information placard at the location, which shows an 1890s photo of the Woodley Lane Bridge and the adjacent Calvert Street trestle.

WILLIAM HOWARD TAFT BRIDGE

The William Howard Taft Bridge, which carries Connecticut Avenue across the Rock Creek Valley, represents the archetype of the Rock Creek and Potomac Parkway bridges, as well as some other bridges in the city. The parkway is where the impressively high-profile bridges are to be found: Taft Bridge is the highest of them. Bridges designed using classical forms, such as Roman aqueduct arches, are found in several locations in the city: Taft Bridge is the earliest of them. Washington, D.C., has many bridges that are decorated with statues, ornate railings, and lampposts: Taft Bridge is the earliest of them, and its lion statues are probably the most widely known.

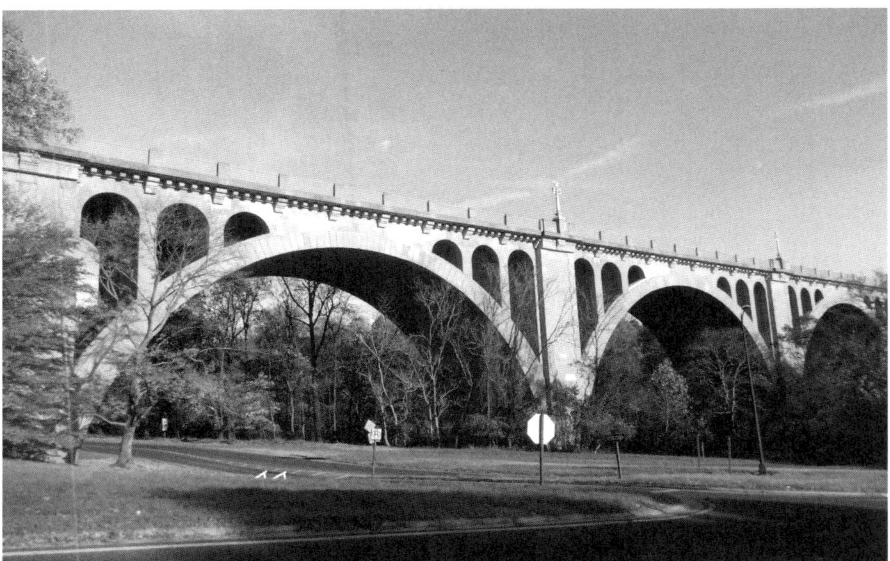

View of the William Howard Taft Bridge carrying Connecticut Avenue across Rock Creek Valley.

High, wide bridges carrying the major thoroughfares were needed to speed commuters across the valley to the newly-opening Northwest D.C. and Maryland suburbs: Taft Bridge was the first of these.

When it was constructed in 1907, the Connecticut Avenue Bridge, later renamed Taft Bridge, was so universally admired that it established the aesthetic model that was used for many later bridges. In 1908, Francis P. Sullivan, writing in the *Architectural Annual*, cited the Taft Bridge and the Piney Branch Bridge (east of Rock Creek Park, discussed in chapter 8) as "modern" examples of how innovative mixtures and surface treatments were being developed to make concrete work appear to look like masonry. In his history of dozens of Washington's bridges in 1938, Emery provided an objective, unembellished subtitle for each new bridge subsection of his report. Key Bridge was discussed under the subtitle "Francis Scott Key Bridge," P Street under the subtitle "P Street Bridge," and so on. However, the Taft Bridge was described under the subtitle "Taft Bridge, A Nation's Pride." All of these factors, as well as the innovative construction method used for the bridge, resulted in its placement on the National Register of Historic Places in 2003.

Of all of the various roadways that facilitated the development of the Washington residential suburbs west of Rock Creek and into Maryland, Connecticut Avenue is probably the most prominent. Most of the fifty state-named "avenues" are internal to the District, and only a handful of them extend, relatively wide and straight, across the Maryland boundary and as far the Capital Beltway. West of Rock Creek, there are only two: Wisconsin Avenue leading from Georgetown to Bethesda, Maryland, and Connecticut Avenue leading from downtown D.C. into Chevy Chase, Maryland. While the other high-profile road bridges across Rock Creek Valley support access to residential areas in Northwest D.C., only Taft Bridge, carrying wide, straight Connecticut Avenue, directly connects downtown to Maryland and the Beltway west of Rock Creek. This makes the Taft Bridge of regional importance, as opposed to the mostly local function of all other Rock Creek bridges, except maybe M Street.

The importance of Connecticut Avenue is only the beginning of the superlatives that can be applied to Taft Bridge. As automobiles were developed and the early, narrow bridges became insufficient for the growing traffic, Woodley Lane near Connecticut Avenue was the first of the bridges to be replaced with a modern, large-scale street bridge. Taft Bridge was constructed between 1897 and 1907 and, at 128 feet high, was the largest unreinforced concrete bridge, with no reinforcing iron bars, in the world.

The seven concrete arches were constructed using an innovative technique. Pre-cast concrete blocks were placed to create the outer facing of the arches, but these then served as forms for cast-in-place concrete pumped behind them. This, along with crushed local rock and sand incorporated into the aggregate, allowed the bridge to appear to be of stone construction. To reduce weight on the arches, the spandrels were left open, with the deck supported by smaller, secondary arches sitting on the main arch. The bridge design received multiple awards and accolades and was used as a model for later open-spandrel bridges, such as Key Bridge.

In addition to reducing weight, the open-spandrel design, especially on high-profile bridges such as Taft, is decorative. The aesthetics of the bridge established the standard that needed to be achieved by all of the later bridges along Rock Creek. The construction between 1897 and 1907 was the peak of the City Beautiful Movement and the work of the McMillan Commission. It also only shortly predated the establishment of the U.S. Commission of Fine Arts, which still controls the aesthetics of bridges and other structures in the District.

Although the form of the Taft Bridge structure dominates the views within the valley for a large distance, drivers and passengers crossing over the bridge see nothing of the enormous arches. These viewers were not forgotten, though, because the sculptures on the corner monuments and the lampposts are among the most ornate in the city. The monuments on the four corners each hold a larger-than-life-sized lion carved out of concrete. The creator was Roland Hinton Perry, one of the major sculptors in the United States during the early twentieth century, a time when large-scale, classically inspired, and allegorical sculptures and reliefs were applied to all types of public structures. The lions are seven feet high and thirteen feet long. Perry's original lions were removed for renovation in 1993 but were too degraded to be restored. Replacement lions were put on the bridge in 2000, with a quarter-sized replica displayed in the lobby of the U.S. Commission of Fine Arts building. The fate of the original lions was unknown until a *Washington Post* reporter found them in 2019. They are currently stored in a side chamber of the tunnel carrying Interstate 395 below Third Street through downtown Washington.

The twelve bronze twenty-two-foot-high lampposts on the bridge are also remarkable sculptures in themselves. Each is a pedestal with draped garlands of bronze fruits, a fluted column topped by a cornice, and then glass globe lights hanging from each side of the cornice. Each lamppost is then topped with a bronze eagle, again larger than life, with spread

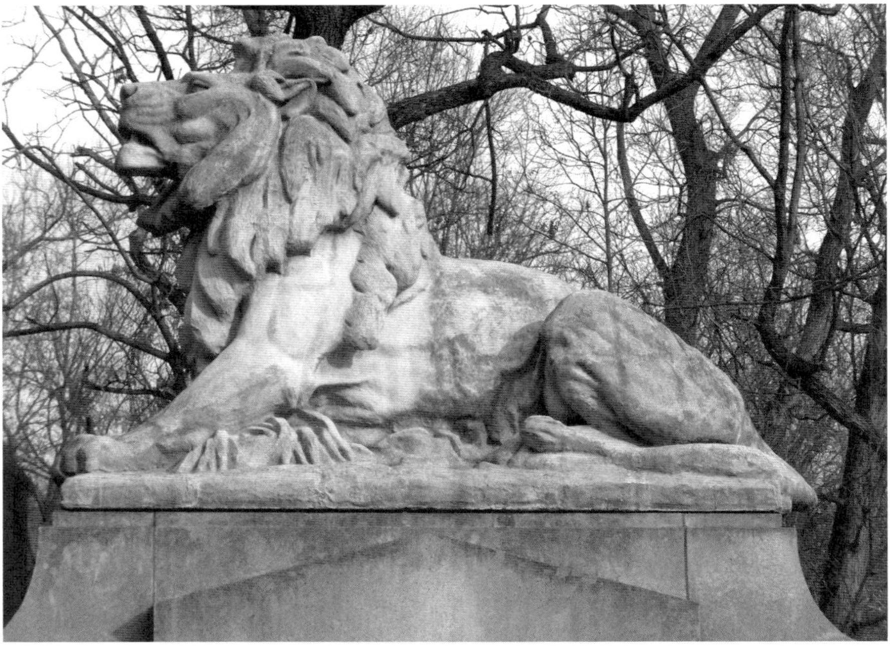

wings. The eagles were designed by Ernest Bairstow, who also sculpted the details on the exterior of the Lincoln Memorial and the lettering for the Gettysburg Address and Second Inaugural Address on the interior of the memorial.

Originally called the Connecticut Avenue Bridge, the high price tag led to it being called the Million Dollar Bridge by the locals. The name of the bridge was eventually changed to honor William Howard Taft in 1931. Taft lived nearby and walked across the bridge every day when he was chief justice of the Supreme Court.

Today, the bridge is a major local landmark, even if most Washingtonians never get to see it. It has sidewalks on both sides, each with amazing views of Rock Creek below. One of the happy circumstances is that the Taft and Ellington Bridges are so close together that each can be used as the perfectly situated viewing and photography platform for the other. There is also a large lawn area on the hillslope adjacent to Shoreham Drive, where local residents can enjoy outdoor activities with an amazing view of the bridge towering above them.

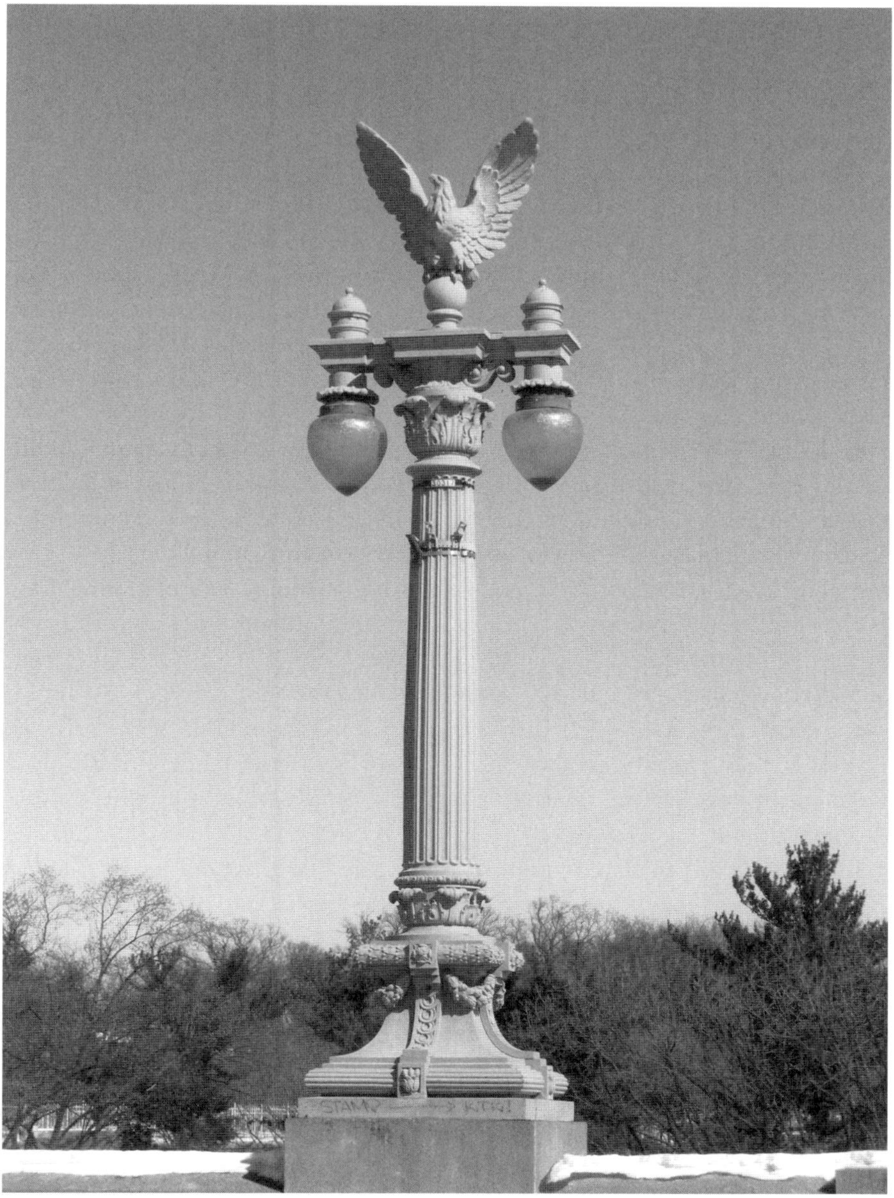

Opposite: One of four lion sculptures on the corners of the William Howard Taft Bridge.

Above: Close-up view of one of the ornate lampposts on Taft Bridge.

CHARLES C. GLOVER (MASSACHUSETTS AVENUE) BRIDGE

To understand the history of bridges at the current location of the Charles C. Glover Bridge at Massachusetts Avenue, it is helpful to review the history of Tiber Creek. At the city's founding, Tiber Creek was a small stream that flowed parallel to Rock Creek, between Rock Creek and the proposed location for the new Capitol Building. Its headwaters were north of the Capitol, and the stream flowed southwest, between the current locations of the White House and the Washington Monument, where it emptied into the Potomac. Early bridges crossed the stream at F Street, E Street, and Pennsylvania Avenue.

Residents of the District today have never even heard of a stream named Tiber Creek, because it was engineered out of existence more than two hundred years ago. Between 1809 and 1815, it was developed into the Washington City Canal, which passed directly through the center of the city from Rock Creek to the Anacostia. By the 1850s, sedimentation had rendered the Washington City Canal unusable, and it stagnated into an open sewer. Beginning in 1871, the canal was filled in, except for the Tiber Creek segment. In between the Capitol and the Potomac, a cover was placed over the canal, which, along with the canal walls, formed an underground conduit for stormwater. Today, this conduit lies underneath Constitution Avenue as it passes along the majestic row of Smithsonian museums on the northern side of the Mall.

This is relevant to the location now occupied by the Charles C. Glover Bridge because, for a few decades near the beginning of the twentieth century, there was a movement to do the same kind of reengineering and development to the southern end of Rock Creek.

In 1888, a steel bridge was constructed across Rock Creek at what is now the location of Massachusetts Avenue. This was the Thompson Bridge, constructed by a realtor wishing to develop farms and fields northwest of Rock Creek into residential properties. Because this was a low-profile bridge, only forty-five feet high, traffic was still required to descend into the valley and then ascend the other side. In 1901, this problem was solved by beginning the process of burying Rock Creek and turning it into an underground culvert, as had been done with Tiber Creek. A brick arch culvert was constructed over Rock Creek, and then fifty-three feet of fill material was brought in to construct a causeway across the valley that was level with the city streets on either side. This allowed vehicles to speed their way across the valley within minutes, thus increasing the value of real estate holdings in Northwest D.C.

Although this first step toward burying the entire southern end of Rock Creek was taken, the idea was eventually displaced by a better one. Instead of burying Rock Creek so that the land area could be developed, Congress authorized the construction of a scenic parkway, now known as the Rock Creek and Potomac Parkway. The parkway was authorized in 1913 and completed in 1936. However, the earth-fill causeway and culvert at Massachusetts Avenue did not conform to this plan. The causeway obliterated the scenic views of the forested hill slopes along the drive from the zoo to Georgetown, views that are a major attraction of the area today. Also, there was no space for a major roadway within the culvert. A narrow road through the culvert was constructed in 1925, but the road flooded frequently and was unsafe due to a narrow curve inside the culvert.

Another development that had occurred since 1901 was the construction of the Taft and Calvert Street (now Duke Ellington) Bridges crossing the entire Rock Creek Valley farther to the north on enormous, visually appealing arches. The Taft Bridge, constructed in 1907, established the model for how to facilitate traffic flow while enhancing the visual appeal of the Rock Creek Valley at the same time. Based on this model, the large earthen causeway at Massachusetts Avenue was removed, and the current Charles C. Glover Bridge was opened in 1941.

The Rock Creek Valley is narrower at Massachusetts Avenue than it is at Connecticut Avenue. The Ellington Bridge requires three arches to span the 760-foot-wide valley, while the Taft Bridge uses seven arches to cross more than 1,300 feet. The valley at Massachusetts Avenue is only 350 feet wide, so the Glover Bridge crosses the entire valley on a single circular span that is 150 feet long. The closed spandrels of the bridge are faced with rough-textured multicolored stone blocks, giving the bridge a rustic appearance that aligns with the objective of enhancing the visual appeal of the parkway. The arch ring is outlined in larger, smooth-textured white blocks. The opening of the arch towers forty feet above the roadway, accomplishing the goal of allowing an uninterrupted view along the axis of the valley.

One of the greatest attractions of the bridge, to tourists and visitors, is the historic architecture and sense of international adventure along Massachusetts Avenue on both ends of the bridge. There is not much to see of the bridge itself, but it is located along the famous Embassy Row. While there are foreign embassies and ambassador residences throughout the city, they are mostly concentrated along Massachusetts Avenue between Dupont

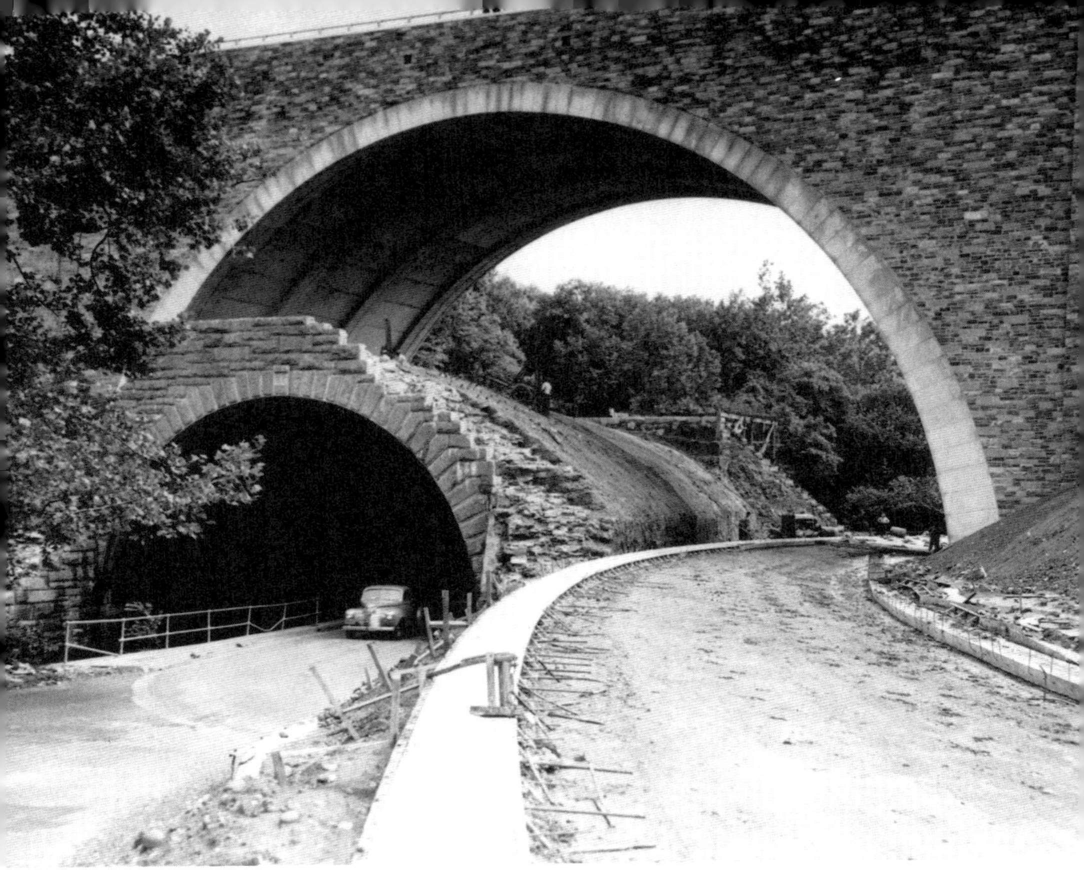

Demolition of Massachusetts Avenue culvert following construction of Glover Bridge in 1941. *Library of Congress.*

Circle on the southeast and the U.S. Naval Observatory on the northwest. Between Dupont Circle and the bridge, you will walk past the embassies of Indonesia, Ireland, Romania, Greece, Latvia, Burkina Faso, Kyrgyzstan, Haiti, Madagascar, Cameroon, Malawi, Ivory Coast, Chad, South Korea, Slovenia, Belize, Japan, Lesotho, India, and Turkey and the ornate Islamic Center of Washington, D.C. Northwest of the bridge to the observatory are the embassies of Italy, Brazil, Bolivia, South Africa, and Great Britain. Dozens more embassies are found on the side streets. Most of these are in historic mansions constructed by American politicians and businessmen and later purchased and converted into embassies.

While strolling and photographing these gorgeous historic buildings, make sure to check out the view of Rock Creek and the parkway, one hundred feet below, while crossing the bridge. The view in both directions is amazing. The high-profile bridges along the parkway are not evenly spaced. The Ellington and Taft Bridges are adjacent to each other just south of the Zoo Tunnel, and Dumbarton, P Street, M Street, and Pennsylvania Avenue come in rapid succession adjacent to Georgetown. However, there is a relatively long gap of approximately one and a half miles in which there is only one bridge, Glover Bridge at Massachusetts Avenue. Due to this distance, some curves in the creek, and the trees, the upstream and downstream views largely do not include any of the other bridges or any other structures. In fall and winter, without any leaves, there is a distant view of Taft Bridge, but the rest of the view is just the creek, parkway, and wooded hillsides.

There are two plaques on the low stone parapet, dedicating the bridge to Charles C. Glover. Glover was a local resident and president of Riggs Bank, one of the city's most prominent financial institutions during the massive redevelopment of the District in the late nineteenth and early twentieth centuries. Glover was an important player in the creation of Rock Creek Park, the National Zoo, and East and West Potomac Parks.

Access to the bridge requires finding on-street parking in the Kalorama neighborhood, which is not easy. There is access to a rustic hiking trail on the southwestern corner of the bridge near the Italian Embassy. The trail leads downhill to creek level within Dumbarton Oaks Park, another sub-unit of the larger Rock Creek Park. Alternatively, the bridge structure can be viewed in profile from the Rock Creek Trail, which parallels the creek and parkway more than one hundred feet below the bridge.

DUMBARTON (Q STREET) BRIDGE

The Dumbarton Bridge, located about a mile southeast of the Glover Bridge at Massachusetts Avenue, may be the most unusual and unexpected bridge in the entire District of Columbia. Like the model established by the Taft Bridge, it is a high concrete arch bridge with a decorative spandrel and sculptures along the deck, but the similarities end there. Compared to Dumbarton, the decorations and sculptures on the other Rock Creek and Potomac Parkway bridges are relatively subtle and muted. Dumbarton takes these decorations to another level, and the difference is apparent as soon as the bridge comes into view around a curve in the parkway.

Also referred to as the Q Street Bridge, Que Street Bridge, and Buffalo Bridge, the Dumbarton Bridge was constructed in 1915. In most cases, decorations on bridges, or any public structure, are intended to commemorate or communicate a theme that is of importance to the community. The decorations may be representative, or symbolic, or allegorical, but there is usually some significance intended by the designer. For the Dumbarton Bridge, the decorations seem somewhat random and appear to have been added just for the sake of decoration. There does not seem to be any other logical explanation for combining a medieval bridge with a Roman aqueduct and an American Wild West theme of Native Americans and bison.

The bridge was designed by Glenn Brown, the executive secretary of the American Institute of Architects, who was one of the driving forces behind the City Beautiful Movement. Brown chose the rounded shape of the five concrete arches based on the Ponte Maggiore in Italy, in the Romanesque Revival style. The arches are seventy-five feet high above the parkway and are reinforced by steel girders enclosed within the concrete. However, they are only about half as high as the entire structure. The upper half of the structure is the most dominant visual feature from a distance and is composed of an intricate row of smaller arches, designed in the manner of Roman aqueducts, with their bases sitting on corbels. These faux aqueduct arches serve to cantilever the sidewalk out over the edge of the deck, allowing more space for traffic lanes.

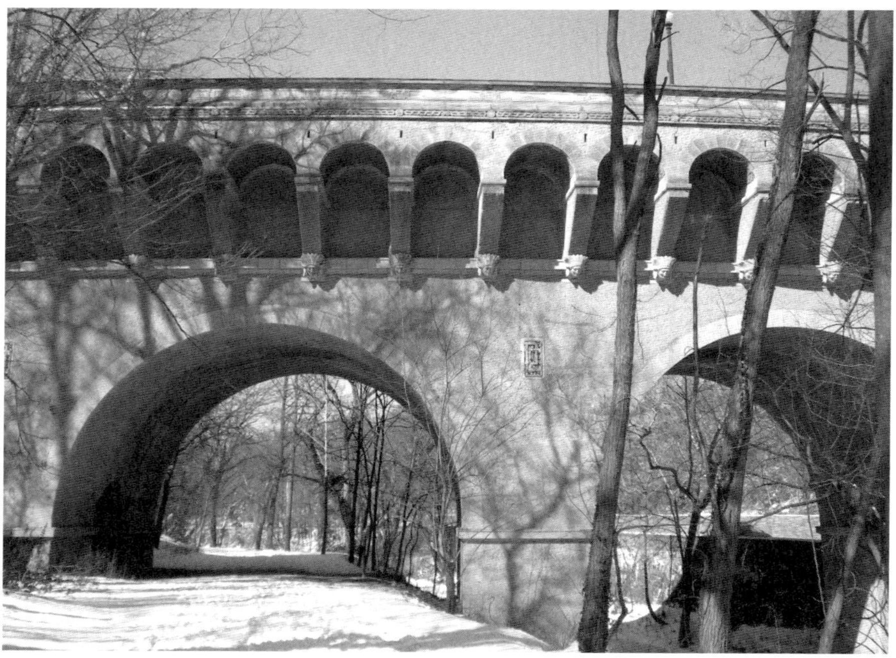

Dumbarton Bridge, with Roman-style aqueduct arches, carved heads of Kicking Bear, and monogrammed initials of "DC."

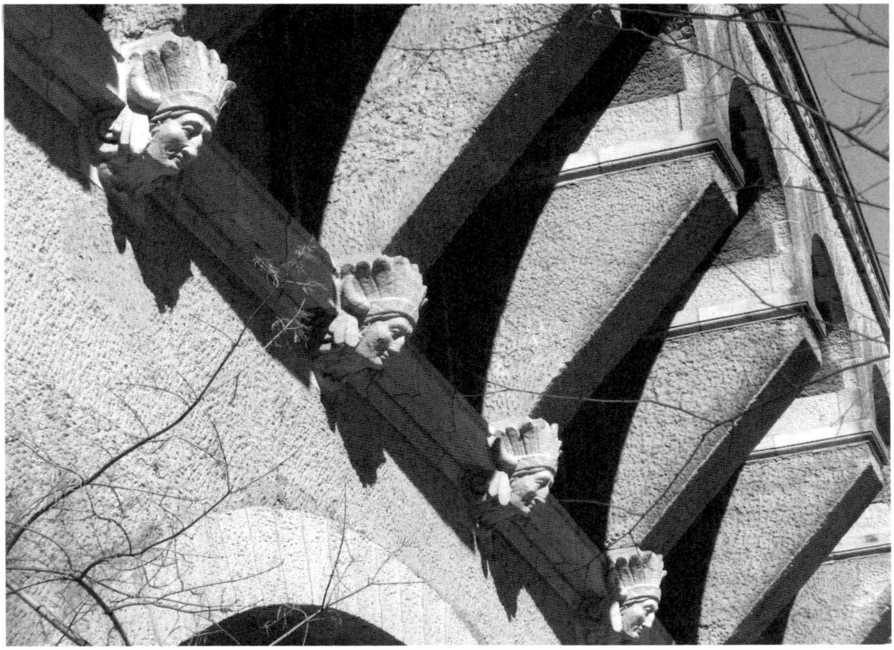

Row of sculptures of Oglala Lakota Sioux Chief Kicking Bear on the Dumbarton Bridge.

Each of the dozens of corbels is carved with a small Native American head, complete with full headdress. The model for the sculpted heads was a life mask of the Oglala Lakota Sioux Chief Kicking Bear. In 1896, Kicking Bear traveled to Washington, D.C., as part of a delegation presenting grievances to the Bureau of Indian Affairs. During this visit, he agreed to have a life mask made for the Smithsonian Museum of Natural History to use as an example of a Sioux warrior. Brown used the mask to create the dozens of sculpted heads used as corbels on the bridge.

Another unusual feature of the bridge is that it is curved—the only bridge in the District that has a curve. In 1895, Congress passed legislation to require that Georgetown streets be renamed to match the corresponding street on the eastern side of Rock Creek. However, this proved problematic, as downtown Washington and Georgetown on opposite sides of the Rock Creek Valley largely developed as separate towns, and except for the bridge at M Street, there was no attempt to make roadways continuous from one side to the other. The streets did not directly match up with each other, and the hilly topography near Rock Creek made a rectangular road pattern in some parts of Georgetown impossible. Thus, when it came time to construct a bridge at Q Street in 1915, the two ends of Q Street were offset by about 185 feet. There would have been many options to connect the two disparate locations with a straight bridge, but Brown elected to enhance the aesthetics of the bridge by making it curved.

Another design choice that makes this bridge distinct from every other bridge in the District is the color and texture. While most of the other Rock Creek bridges use multicolored stone facing or concrete to achieve a general shade of gray, the Dumbarton Bridge is a reddish buff color, intended to mimic terra-cotta and evoke the feeling of a Mediterranean setting. This was achieved using Kingswood sandstone from West Virginia for the carved stone decorations and mixing different types of sand and gravel to mimic the same color for the concrete. The bridge appears to be orangish, or pinkish orange, depending on the sun. The concrete surfaces were also textured with a rough, stippled pattern to appear similar to a terra-cotta structure. Barely noticeable, unless you know to look for them, are intricately carved monograms of the initials "DC" on small panels on the spandrel in between the main arches.

Similar to the other parkway bridges, the elaborate decorations on the sides of the bridge can be seen only briefly by passengers in speeding cars

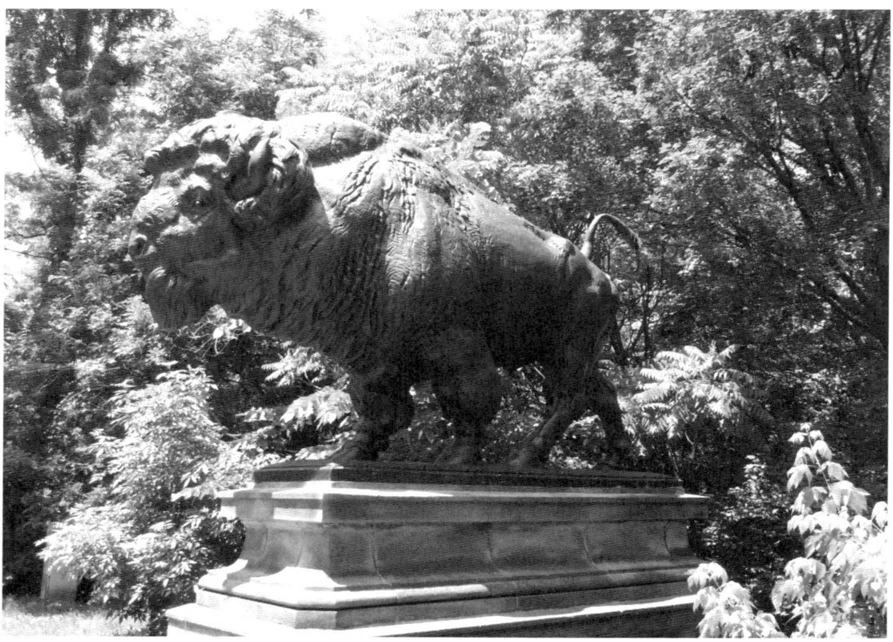

One of four life-sized bronze bison on the corners of the Dumbarton Bridge.

driving underneath the bridge. However, they can be viewed at leisure by hikers and cyclists on the Rock Creek Trail adjacent to the parkway, and there are decorations on the deck and along the sidewalks as well. The parapets are sandstone carved into a decorative series of geometric panels connected by diamond shapes. At the ends of the parapets, on all four corners, are monuments topped by enormous, life-sized bronzes of bison by prominent sculptor Alexander Phimister Proctor. These have resulted in the popular local nickname for the bridge, Buffalo Bridge. These unusual construction methods and elaborate decorations were the basis for listing the bridge on the National Register of Historic Places in 1973.

Q Street is a residential neighborhood on both ends of the bridge, and the bridge has sidewalks on both sides. There is no dedicated parking or close Metro access, but on-street parking can be found.

P STREET AND NEAR P STREET BRIDGES

Almost immediately on emerging from beneath the unusual Roman-style aqueduct lined with carvings of the heads of Native American chiefs, Rock Creek and Potomac Parkway crosses from east to west over the creek on the "Near P Street" Bridge, which does not actually carry P Street, and then plunges beneath the arch of the P Street Bridge, which does.

The location of the P Street and Near P Street Bridges is one of the most historically important crossings of Rock Creek. At the time of the city's founding, Rock Creek was tidal up to this location, making this the natural place for early travelers to ford the creek. The ford played a significant role in the Revolutionary War. After the early battles in New England, New York, and New Jersey, the British sailed south of the Chesapeake Bay in 1780 to carry the war into Virginia. The Continental army, supported by French troops under the Marquis de Lafayette and Comte de Rochambeau, followed on foot, gathering additional recruits along the way. In April 1781, Lafayette used this ford to cross Rock Creek, followed by the Baltimore Light Dragoons in June and the Duc de Lauzun's Legion of cavalry in September. Shortly after, Lauzun became one of the heroes of the American and French victory at Yorktown, which ended the Revolutionary War. The following summer, Rochambeau's army recrossed at this ford on their way back north.

The earliest bridge at the P Street location was the wooden covered Paper Mill Bridge, constructed in about 1855 and named after a mill on the eastern side of the creek. The Paper Mill Bridge was a low-level bridge, crossing Rock Creek at an angle, from north to south. It was replaced with a steel Pratt truss bridge in 1871, but that bridge required major reconstruction in 1893 and 1923.

The low-profile 1923 bridge was eventually removed in 1934 to make way for two replacement bridges: one low-profile bridge to carry the new Rock Creek and Potomac Parkway on a north–south orientation across Rock Creek and a high-profile bridge to carry P Street on an east–west orientation across both Rock Creek and the parkway.

The low-profile bridge, now known as the Near P Street Bridge (yes, that is the real name!), was opened in 1936, completing the full length of the parkway. It is a reinforced-concrete arch bridge faced with multicolored blocks of mica schist and a decorative arch ring, conforming to the NPS practice of giving park bridges a rustic appearance. Interestingly, the 1988 Historic American Engineering Record for the bridge states that it has sidewalks on both sides, which is no longer the case. It is likely that post-1988 improvements to the Rock Creek Trail on the western streambank made a sidewalk unnecessary.

The "real" P Street Bridge, also known as Lauzun's Legion Bridge, hovers almost directly above the Near P Street Bridge. Funded by the PWA and opened in 1935, it is a two-span, closed-spandrel elliptical concrete arch, faced with multicolored stones. The arch ring is composed of larger white blocks and the keystones are oversized, reaching all the way to the parapet. The parkway passes through the western arch, while the creek passes through the eastern. The two sidewalks are lined with plain, but attractive, granite parapets. There is a bronze plaque with a green patina on either end of the parapet, identifying the commissioners responsible for the bridge in 1933 and 1934.

With the flatter topography and elevation close to sea level, the P Street Bridge is only about forty-five feet high, in contrast to the more than 120-foot-high spans farther upstream at the Taft and Ellington Bridges. The bridge crosses the entire valley, including the creek and the parkway, marking the transition from the impressive residential townhouses on P Street in Georgetown on the west to the urban apartment and office buildings of downtown to the east.

The history of the two bridges reveals the personal interest that Spratt took in describing the bridges of Rock Creek in 1953. In 1934, the Daughters of the Revolution placed a bronze plaque on the Near P Street Bridge, commemorating the use of the location by Lauzun's Legion. Spratt, in his report to the Columbia Historical Society, took umbrage at the placement of this plaque, pointing out that the colonial road used by Lauzun's Legions crossed the ford on a direct east–west alignment, as opposed to the north–south alignment of the Paper Mill and Near P Street Bridges. Clearly disturbed by this injustice, Spratt argued, in what was otherwise a dispassionate technical report, that the bridges be renamed and the plaque moved to honor the correctly aligned P Street Bridge instead of the misaligned Near P Street Bridge. While this story sounds as if it is just historical entertainment, the District apparently agreed. More than fifty years later, in 2006, the Lauzun's Legion name was transferred to the P Street Bridge.

The deck of the P Street Bridge can be accessed by using on-street parking either in Georgetown on the western end or in downtown on the eastern end. The bridge can be viewed in profile by using the Rock Creek Trail on the western end, but a more interesting photo opportunity is found on the eastern end. At the southeastern corner of the bridge at the intersection of P Street and Twenty-Third Street, a dilapidated set of stairs leads down to P Street Beach, an urban park on the eastern bank of the creek between P Street and M Street. Most of this park is to the south of the P Street Bridge, but a part of the trail leads back north underneath the eastern arch of the bridge. From this trail, there is a remarkable view of three bridges—P Street, Near P Street, and Dumbarton—one after another. The visual appeal comes from not only the outer decorative treatment of the three different bridges but also their different heights and the different angles at which they cross the creek.

M STREET BRIDGE

The Georgetown neighborhood of the District extends almost a mile from north to south and more than a mile from west to east. The western portion of this area is occupied by Georgetown University, and the northeastern portion is occupied by Dumbarton Oaks Gardens, Dumbarton Oaks Park, and Montrose Park. Most of the rest, north of M Street, is historic townhomes on residential streets. However, none of this development existed in the 1780s and 1790s, when the tobacco port town was included within the boundaries of George Washington's new Federal City.

The historic core of Georgetown, as depicted in a plat developed when the town was chartered and named George-Town by the Maryland

Assembly in 1751, was an intersection of four major roadways situated in the middle of a narrow strip of land along the Potomac shoreline. High Street, a common name for a main market or shopping area, extended uphill directly north from this intersection. The appropriately named Water Street extended directly south to the Potomac shoreline. Today, these are called Wisconsin Avenue. Extending in the direction of Little Falls to the west, Falls Street is also as much of a descriptor as a name. And extending east of the intersection, its name giving a hint of what may have lain off of the right-hand side of the plat in 1751, is Bridge Street. Today, Falls Street and Bridge Street are known as M Street, the main east–west thoroughfare that tourists and visitors tend to think of as "Georgetown." It is where the shops, restaurants, clubs, and hotels are to be found.

When researching the history of bridges in Washington, D.C., it is easy to find an unchallenged, commonly accepted location and date for the very first bridge within the boundaries of the District. It was a wooden bridge constructed at the location where today's M Street, yesteryear's Bridge Street, crosses Rock Creek. The location makes sense. Georgetown was, by far, the largest and most prosperous settlement, and Rock Creek, on Georgetown's eastern boundary, was big enough to require a bridge but small enough to easily be bridged. There also does not seem to be any dispute regarding the construction date of this first bridge. All sources agree that it was constructed by the Town of Georgetown in 1788, two years before the area was selected as the new Federal City.

Interestingly, 1788 would be a full thirty-seven years after the road leading to that location was depicted on a map as "Bridge Street." According to the *Georgetown Historic Waterfront, A Review of Canal and Riverside Architecture*, published in 1968 by the U.S. Commission of Fine Arts and National Park Service, Bridge Street led to a "portage" over Rock Creek, creating even more confusion. By definition, a portage is an overland trail used to bypass an unnavigable section of stream or river. It is not a bridge, ford, or ferry. In both 1751 and 1788, Rock Creek was navigable all the way up to what is now P Street and was regularly used by ships. The use of the word *portage* does not seem to accurately describe the situation.

Instead, while the history books insist that the date of the first bridge in the District was 1788, it seems that there is something missing from the story. It may be more plausible that a bridge existed at the Bridge Street location in, or shortly after, 1751. Technology existed for wooden, stone, and moveable bridges at that time, so there would be no reason an earlier bridge could not have been constructed. We know that there was no bridge there in 1781, because Lauzun's Legion crossed the creek on a ford at P Street. However, that does not preclude a bridge having been constructed in 1751 or earlier. These early bridges were routinely destroyed and reconstructed.

Moving on from the question of whether the first bridge was constructed on Bridge Street prior to 1751 or in 1788, there is general agreement that the 1788 bridge collapsed under the weight of a stagecoach a few years later, drowning the driver and four horses. It was reconstructed in 1800, with a drawspan so that ships could continue to sail as far inland as what is now P Street. In 1839, this bridge was replaced with a wooden covered bridge, which lasted until 1871. At that time, it was replaced with a deck truss steel bridge. The 1871 bridge was condemned in 1925, and the current bridge opened to traffic in 1930.

The design of the current M Street Bridge was not without controversy. It was designed and constructed during the same period as the rest of the Rock Creek and Potomac Parkway but departed from the usual "parkitecture" style favored by the U.S. Commission of Fine Arts for the other bridges on the parkway. Of four low-profile bridges and eight high-profile bridges crossing the creek south of the zoo, M Street is the only one that is not a concrete arch and not faced with stonework. Instead, it is a flat steel-girder bridge, although it was dressed up to hide the girders and appear more decorative. The authorizing legislation from Congress required that the District commissioners "consult" with the Commission of Fine Arts—in their opinion, they did. Never mind that the Commission of Fine Arts angrily rejected the design every time it was presented, referring to it as an "ugly service bridge unfitted for park purposes." The Commission of Fine Arts favored another concrete arch faced with stone to match all of the other bridges on the parkway.

Despite this departure from the design of the other bridges, the current bridge does not stand out as an aesthetic mistake, for two main reasons. First, the applied decoration is not unappealing. The concrete piers are formed to appear to be stonework and have small decorative arches in them at road level. The steel girders are faced with concrete formed in a series of rectangular panels. The sidewalks are cantilevered out over the edges of the deck, sitting on corbels, a design that is also used on the Dumbarton Bridge. Neoclassical balustrades line the deck. The only unattractive part is that the underside of the deck is not masked, revealing that, in addition to a roadway, the bridge carries an industrial-looking network of conduits and a forty-eight-inch water main.

The second reason is that this southern section of the parkway is far less "parklike," and far more urban, than the northern section, so the more utilitarian-looking bridge design is not discordant with its surroundings. The Rock Creek and Potomac Parkway here is far less of a parkway and more of a busy urban four-lane highway. The M Street Bridge marks the end of the section of the parkway that is lined with tree-covered hillsides. The streambanks of Rock Creek north of M Street are occupied by greenspace of Rose Park on the west and P Street Beach on the east. However, the M Street Bridge foreshadows that the parkway is about the enter an unattractive, complex network of on- and off-ramps for K Street and the Whitehurst Freeway. Although the "official" Rock Creek and Potomac Parkway continues along the eastern shore of the Potomac River south to the Lincoln Memorial, the quiet rusticity of Rock Creek Park has been left far behind.

PENNSYLVANIA AVENUE BRIDGE

In outward appearance, the Pennsylvania Avenue Bridge across Rock Creek and the parkway does not look any different from, and seems as if it would have a similar history to, the other bridges along the parkway. However, this is not the case. The stone-faced concrete arch of the bridge is hiding an unusual secret, one that provides insight into the early growth of the city.

The impetus behind the construction of the Washington Aqueduct, which began in the 1850s and was supervised by General Montgomery Meigs, is discussed in chapter 4. The Washington Aqueduct was a system of reservoirs connected by open conduits and pipes, depending on the terrain. The water intake for the system was at Great Falls, and the aqueduct followed the Potomac Palisades on the northeastern shoreline of the river to Georgetown, downtown, and the Navy Yard. On its route, the aqueduct used bridges to cross six substantial tributaries entering the Potomac River from the east. One of these was at Cabin John Creek, which was crossed by the aqueduct on the Union Arch Bridge. Another was at Rock Creek, where the water was transported from Georgetown into downtown.

In the 1850s, Pennsylvania Avenue, which was the major northwest–southeast thoroughfare in the city, terminated before reaching Rock Creek. Although the 1791 L'Enfant Plan showed Pennsylvania Avenue crossing Rock Creek on a northwest-to-southeast-oriented bridge to meet with

Bridge Street in Georgetown, that bridge was not constructed. Instead, an east-to-west-oriented bridge was constructed only a few hundred feet to the north at M Street. Because the terminus of the Washington Aqueduct was to be at the Capitol and Navy Yard southeast of Georgetown, the route of the pipeline was designed to follow Pennsylvania Avenue directly southeast, instead of following M Street directly east. Meigs placed this bridge, known as Bridge 6, at the spot where the diagonal route of Pennsylvania Avenue intersected Rock Creek.

The Washington Aqueduct at Pennsylvania Avenue was not originally intended to serve as a traffic bridge. With an existing traffic bridge a few hundred feet north, the only need was for the two forty-eight-inch-diameter water pipes to cross the creek. To do this, Meigs designed the two pipes in the shape of an arch, two hundred feet long, so that the pipes would be self-supporting as they crossed the creek. The pipes were anchored on each end by large stone abutments.

Construction started in 1858, but the plan was soon changed to include a roadway, completing L'Enfant's design to provide a direct connection along Pennsylvania Avenue from Georgetown to the President's Mansion and the Capitol. An iron girder framework and wooden deck was constructed on top of the weight-supporting arched pipes. The resulting structure is shown in historic photographs from the 1860s and is more decorative and ornate than might be expected. The spandrel above the arch is open, with the girders in a geometric X-shaped pattern. The stonework on the abutments is decorative, as is the iron railing. The only part that looks industrial is the two arched

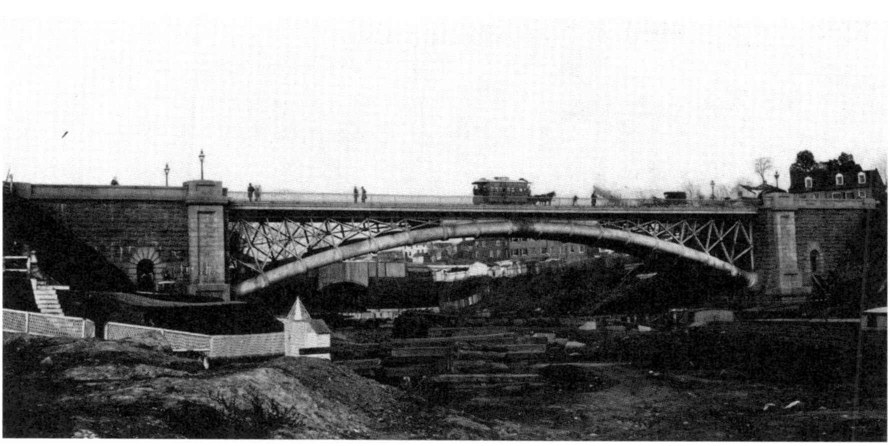

Bridge constructed of water pipes carrying the Washington Aqueduct across Rock Creek at Pennsylvania Avenue. *Library of Congress.*

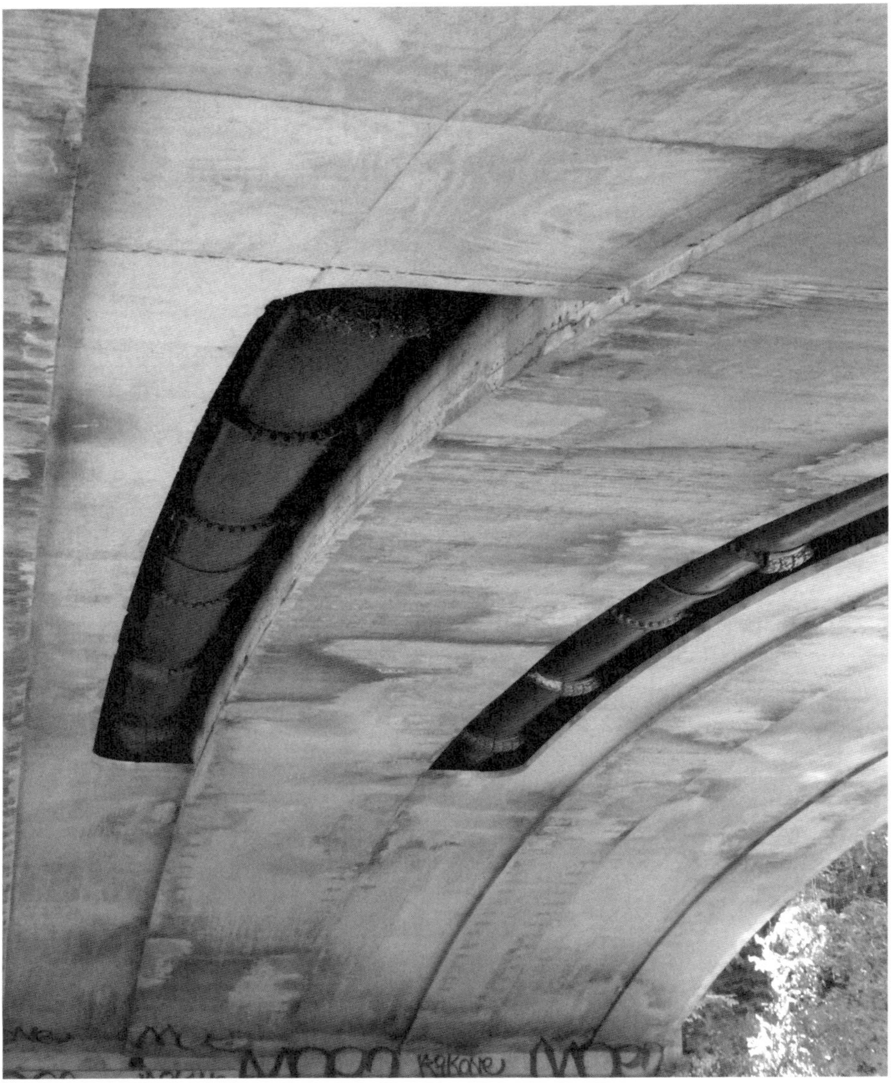

View of the original 1858 water pipes of the Washington Aqueduct, exposed beneath the arch of the Pennsylvania Avenue Bridge.

water pipes. The bridge was recognized for its unique engineering, as well as its aesthetic qualities.

Remarkably, the 1858 water-pipe bridge is still in place and still carries water into downtown. It is in one of the busiest parts of the city, and thousands drive over it every day without even knowing anything about the history or the unique structure. This is because the water-pipe structure was enclosed within a larger, wider concrete arch structure in 1916. According to the *Streets of Washington* blog, the water pipes are no longer weight-bearing, that task being provided by the concrete arch. However, they can be seen, if you know where to go to look for them.

The need for a larger concrete bridge at the location was not due to impending failure of the existing water pipe bridge. It was because the existing bridge was too narrow for the increasing traffic and needed to be widened. The U.S. Commission of Fine Arts, which was established in 1910, successfully fought proposals to replace the entire structure with a steel girder bridge, opting instead for an arch. The current bridge, completed in 1916, generally matches the stone-faced concrete arch motif of the later bridges on the parkway, including Ellington, Glover, and P Street.

Because the bridge is only a few hundred feet from M Street, both bridges can be investigated in a single, short hike from the surface streets in Georgetown. Other than rectangular panels carved onto the granite parapet, there is no substantial decoration along the deck, and no major landmarks may be found at its ends. The underside of the deck can also be studied from the Rock Creek Trail beneath the bridge, where gaps have been left in the concrete to reveal the 160-year-old pipes, still carrying water into the city.

K STREET BRIDGE

The early years of attempts to establish bridges across Rock Creek and connect Georgetown to the new Federal City were apparently a bit chaotic, and the available documentation does not agree on the sequence of events. What is generally accepted is that a wooden bridge at Bridge Street (now M Street) was constructed in 1788, two years before the area was selected as the location of the Federal City. It also seems to be accepted that this wooden bridge collapsed about four years later, which would be 1792.

Instead of reconstructing the bridge at the same location, construction began on a new bridge, situated at what is now K Street, on July 4, 1792. Because this bridge was closer to the river than Bridge Street, the site was mostly marshland. The construction required filling and draining of the marsh, resulting in the street being named Causeway Street. The structure was mostly a causeway, with three stone arches spanning the mouth of Rock Creek.

By all reports, the bridge was handsome, fitting for the capital city of the new nation. George Washington visited the construction site. The center arch was constructed of thirteen stones, representing the thirteen original states, with Pennsylvania as the keystone, resulting in the nickname for the commonwealth that is still used today. The structure included a quay, constructed of Aquia Creek sandstone, which was used as a landing for prominent persons arriving in the city by ship.

After this, the documentation diverges. Spratt reports that the Causeway Street Bridge was pronounced unsafe, condemned, and removed about four years after it was constructed. This would be 1796 or 1797. The National Park Service Historic American Engineering Record, written in 1992, provides a date of 1795, stating that the center arch was "taken down." Both also report that the bridge was not rebuilt until 1869. Instead, the government returned to the Bridge Street location and constructed a new wooden bridge, with a drawspan, in about 1800.

In contrast, other authors, including Emery in 1938 and Van Dyne and Lewis more recently, tell a different story. Emery and Van Dyne relate that the three-arched bridge, with the names of the thirteen original states engraved on the stones, was in place on June 4, 1800, when John Adams made his grand entry into the capital across the bridge as president. Lewis goes further, stating that the stone arch collapsed shortly after it was constructed but that it was reconstructed and crossed by Adams in 1800.

This was not an unimportant visit. This was the formal transfer of the capital of the nation from Philadelphia to the new Federal City, which had been under construction for ten years. On May 15, 1800, Adams had ordered that the entire apparatus of the federal government close up shop in Philadelphia and be prepared to reopen for business in Washington, D.C., on June 15. Adams himself, who had been president for more than three years by that time, had never traveled south of Pennsylvania. Lewis even provides a detailed description of the specific route taken by Adams from Georgetown, directly across the bridge at what is now K Street to Washington Circle, then bearing right on what is now Pennsylvania Avenue to the President's Mansion. Van Dyne similarly stressed the symbolic importance of the role played by the ornate stone arch bridge on that day.

However, this series of events could not have occurred if the sequence described by Spratt and the National Park Service is accurate. Of course, the symbolic transfer of the government from Philadelphia to Washington across an ornate stone arch bridge celebrating the original thirteen states makes for a better story than an entry across a wooden drawbridge. This is speculation, but it may be worth considering the additional facts that in June 1800, the Northern states were still trying to stop the transfer of the government to Washington and that Adams was engaged in a contentious reelection campaign, which he would eventually lose to Thomas Jefferson a few months later. Given that few people in the United States at that time were ever going to see this bridge, there may have been some political benefit in embellishing the story of Adams's grand entry into the new capital city over a prominent bridge that commemorated the original thirteen states.

Like the other Rock Creek bridges, the bridge at K Street went through several different generations. The 1869 version was constructed of wood. This was replaced by a steel-plate girder bridge in 1907.

After the completion of the Rock Creek and Potomac Parkway in 1936, traffic volumes increased to the extent that another replacement was needed. The bridge crossed the creek but carried K Street at the same elevation as the parkway, requiring an intersection. The current bridge, opened in 1941, was designed to eliminate this intersection by crossing both the creek and the parkway, creating grade separation and improving the flow of traffic. Consistent with the other parkway bridges, the U.S. Commission of Fine Arts requested that the bridge be a concrete arch, faced with multicolored gneiss.

Almost immediately upon the bridge's completion in 1941, plans were developed for an elevated riverside freeway that would connect Key Bridge to the Rock Creek and Potomac Parkway. The purpose of the Whitehurst Freeway, which was completed in 1949, was to reduce through-traffic on M Street in Georgetown by providing a bypass that had no intersections. The freeway, along with an unattractive network of on-ramps and a later connection to Interstate 66, was constructed directly above the K Street Bridge. Visually, the complex interchange is unfortunate, not just at the Rock Creek crossing but along the length of the elevated freeway. By being elevated, the freeway completely obliterates any visual connection between historic Georgetown and the waterfront along the Potomac.

8
ROCK CREEK TRIBUTARIES

Small tributaries enter Rock Creek from the east and the west and are crossed by both road and footbridges. Some of these are within the main part of Rock Creek Park, while others are in smaller subunits. The tributaries include Broad Branch, Klingle Stream, Pinehurst Branch, and The Branch entering Rock Creek from the west and Fenwick Branch and Piney Branch entering from the east.

Most of the bridges associated with the tributaries are small wooden footbridges carrying hiking trails or low-profile bridges carrying local roads. However, two of the tributaries, Klingle Stream and Piney Branch, flow within deep, steep-sided valleys and are crossed by major roadways, resulting in decorative, high-profile bridges.

KLINGLE VALLEY BRIDGE

Although the Klingle Valley Bridge is almost a mile away, its history is intricately linked with that of the Taft and Ellington Bridges. Klingle Valley extends about three-quarters of a mile west of Rock Creek, separating the Peirce-Klingle estate on the north from the zoo to the south. The valley is crossed by Connecticut Avenue about three-quarters of a mile northwest of the Taft and Ellington Bridges.

Today, the Klingle Valley and Taft Bridges line up neatly along a wide, straight, major roadway. However, they were not originally constructed that way. Instead, the Woodley Lane Bridge, constructed in 1889, crossed Rock Creek at a different angle than the later Taft Bridge. It was a low-profile bridge that required traffic to descend the valley slope on one side and then ascend on the other, with no connection to the current route of Connecticut Avenue.

Separately, the original Calvert Street and Klingle Valley Bridges were constructed two years later by Francis Newlands's Rock Creek Railway, specifically to carry streetcars to open up the Northwest D.C. and Maryland suburbs to residential development. The Rock Creek Railway approached Rock Creek from the east, crossed on the Calvert Street Bridge, and then turned right to proceed northwest along what is now the route of Connecticut Avenue. The streetcar then crossed Klingle Valley north of the zoo on the original Klingle Valley Bridge.

After completion of the Connecticut Avenue Bridge (now Taft Bridge) in 1907, use of the Klingle Valley Bridge became complicated. By replacing the Woodley Lane Bridge, the Connecticut Avenue Bridge completed a continuous roadway that was straight, wide, and at a constant elevation across Rock Creek Valley. As automobile traffic increased, Connecticut Avenue became a major thoroughfare, providing access between downtown and the Maryland suburbs. However, with the increased volume of automobile traffic on the streetcar-less Taft Bridge, the old, narrow Klingle Valley Bridge, designed for streetcars, became a hindrance to the smooth flow of traffic. It needed to be replaced with a wider modern bridge. The current Klingle Valley Bridge was constructed in 1932. Streetcars stopped using the bridge in 1935, a result of the gradual replacement of trolleys with buses, but the tracks were not removed from the bridge until 1977.

The Klingle Valley Bridge is one of only six bridges in the entire District that serves as prominent public art, with elaborate decorations, sculptural lampposts, and geometrically patterned railings or parapets. Although it does not have statues, its railings are probably the most elaborate in the city. They are constructed of wrought-iron grids inset with cast-iron panels set in an art deco chevron pattern. The bridge is famous for its large decorative bronze lanterns in the shape of urns, positioned on the four corners, as well as tall, ornate lampposts that once served as supports for the cables that powered the streetcars. By the time the bridge was listed on the National Register of Historic Places in 2004, the lanterns and lampposts had severely degraded. However, they were restored during

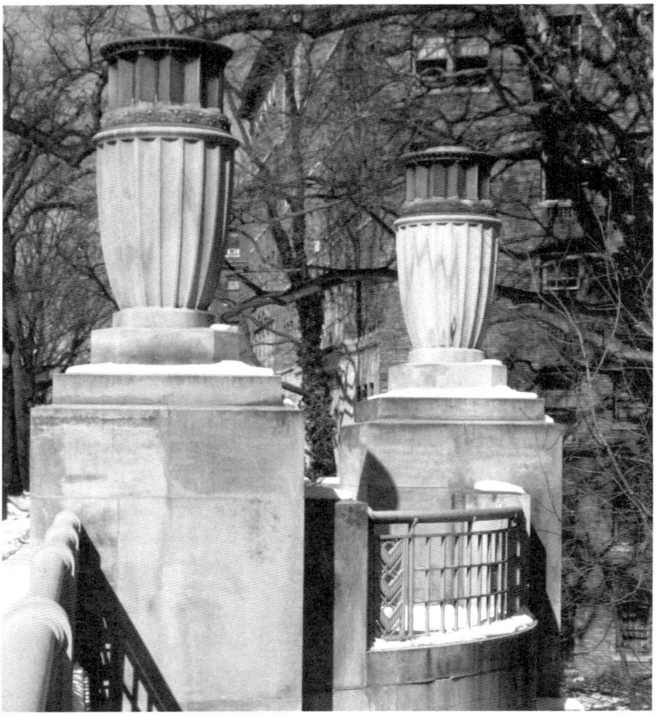

Left: Lampposts in the shape of urns at the corners of the Klingle Valley Bridge on Connecticut Avenue.

Below: View of the 1932 Klingle Valley Bridge carrying Connecticut Avenue across the Klingle Valley.

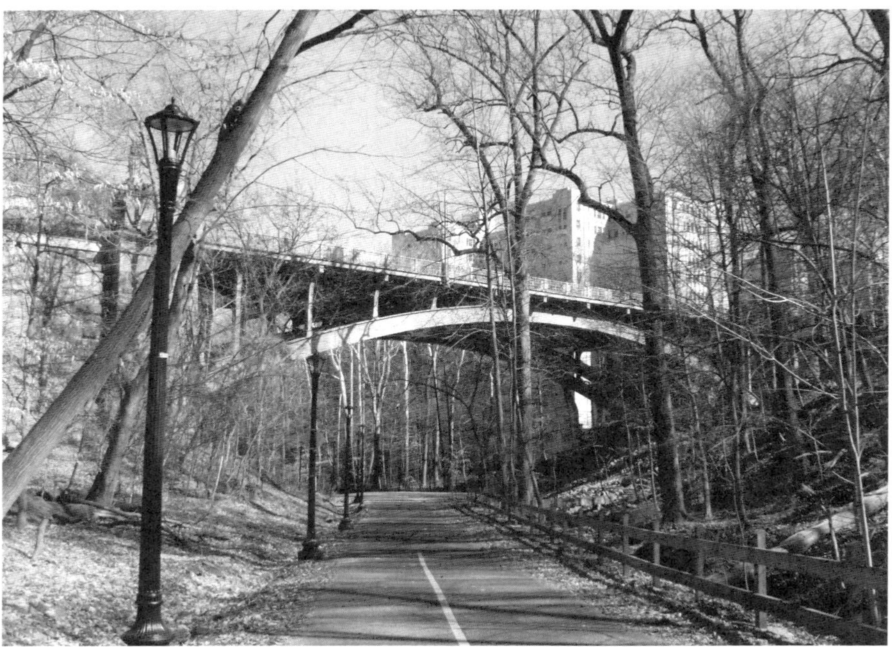

the bridge's rehabilitation in 2005 and remain a major landmark of the Cleveland Park neighborhood.

Klingle Valley and the Klingle Stream that flows through it derive their name from the Klingle family that was associated with Peirce Mill beginning in 1869. Klingle Road was a historic street that ran from the Washington National Cathedral on the west, through Klingle Valley, and ended near Beach Drive just south of the Peirce-Klingle estate.

In 1991, the roadway within Klingle Valley was destroyed by flooding and remained closed for more than twenty-five years. In 2017, the section of the road within the valley was converted to a paved and landscaped multiuse trail that passes directly beneath the high and historic bridge. There is no view of the bridge from adjacent roads or bridges, so this trail is the only place to view the bridge's profile. There is no vehicle access to the trail and, like Taft Bridge, no convenient way to walk from the deck of the bridge to the trail below. Because the trail is so steep and has no parking spaces on either end, bicyclists mostly do not use it, limiting access to dedicated hikers willing to climb the precipitous path from the Rock Creek Trail.

GRANT ROAD BRIDGE AND BROAD BRANCH FOOTBRIDGE

Broad Branch and Broad Branch Road form part of the western boundary of Rock Creek Park between Military Road and Blagden Avenue, near the Nature Center and Planetarium. Broad Branch is a minor narrow tributary that nonetheless is crossed by two interesting old stone bridges. There is a street bridge carrying Grant Road across Broad Branch and a stone pedestrian bridge crossing the creek just west of the park boundary.

The Grant Road Bridge is nothing more than an elaborate stone-arched culvert at the corner of Grant Road and Broad Branch Road. The Historic American Engineering Record for the bridge gives its construction date as circa 1898 but speculates that it could be older, predating the establishment of Rock Creek Park. While the culvert itself is not listed on the National Register of Historic Places, it is discussed as a contributing feature to the Rock Creek Park Historic District.

The twenty-foot-long culvert is almost unnoticeable to passing traffic, but it is not without some decorative stonework. The part of the structure that is visible to drivers and passengers is a low stone parapet, no more than two feet high and hidden behind overgrown vegetation during some parts of the year. It is curved, like a wing wall, to allow for turning traffic. The ends of the wing wall are topped with a capstone that has been roughly shaped with geometric grooves. There is no parking available near the culvert, and Broad Branch Road is a narrow, winding two-lane road with no shoulder. The culvert is also not near any bike or hiking trails. There is space, barely large enough for a single car to pull over along Broad Branch Road, but not much to see.

About a half mile northwest of Grant Road, near Twenty-Seventh Street, is an odd, unexplained stone footbridge in the woods, just a few feet off the northern side of Broad Branch Road. The area is entirely wooded and not far from some isolated residences. It is not located within the boundaries of Rock Creek Park or any other park. Judging by its tiny size of less than five feet wide, the bridge was never anything more than a footbridge, but there is no evidence of a formal hiking trail on either end of it.

Small, decorative footbridge crossing Broad Branch.

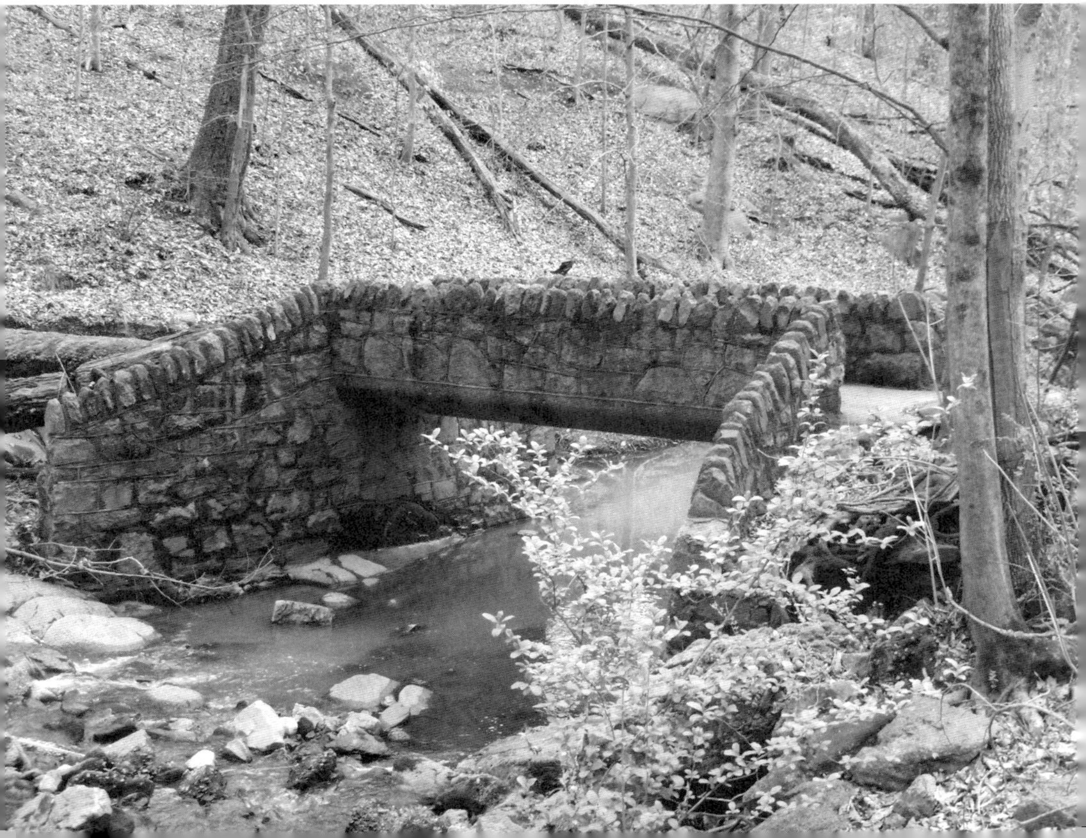

What is odd about the bridge is that, despite no available documentation about its construction or use, the decorative stonework shows that it was clearly a special bridge at one time. The bridge is not a stone arch. About ten feet long, it consists of a flat concrete slab held between two gray stone abutments and lined with stone parapets about two feet high. For no apparent reason, the abutments flare out on either end into wing walls, even though the bridge is far too narrow to have ever carried motorized traffic. A course of the gray stones, decoratively set upright on their ends, lines the top of the wing walls and stone parapets. The fact that the stone parapets, abutments, and wing walls adorn a completely flat bridge suggests they were purely decorative. If they had been structural, they would have had to form an arch.

Like Grant Road, there is no dedicated parking or formal shoulder on the road, but a single car can fit into a small space off the road to allow visitors to investigate. The northern end of the bridge, opposite the road, is damaged by flood erosion, but it does not seem to matter because there is no road or hiking trail on that end. There is also no evidence of a hiking trail on the opposite side of the road or along the road, and a review of maps of the area does not show any formally recognized hiking trails in the area. But someone, at one time, took great pride in constructing this charming little bridge, and it is a shame there is so little information available.

ROSS DRIVE BRIDGE

Ross Drive is an unusual roadway and may be one of the District's least known and most unexpected roads. At the time of its formal designation and subsequent acquisition of the land in 1891, the road network in Rock Creek Park was an ad hoc mixture of roads developed to support farms, milling operations, and the military. In 1897, the Board of Control of Rock Creek Park developed an integrated road network designed to provide access into the park for urban residents and recreational users.

The major roadway developed between 1897 and 1912 was Beach Drive, which parallels the creek from the Maryland boundary to south of the zoo. East-to-west-oriented spur roads such as Morrow Drive, Glover Road, and Wise Road were also constructed at that time to provide access to the street network outside the park. These spur roads linked Beach Drive to Sixteenth Street on the east or Oregon Avenue on the west.

The program also included Ross Drive, which did not connect anything to anywhere. It did not lead anywhere in the early 1900s, and it does not lead anywhere today. It is nothing but a quiet, winding, narrow drive through the woods on the western side of Rock Creek Park. Oriented north–south, parallel to the creek and Beach Drive, it is situated about a quarter mile to the west, about halfway up the western slope of Rock Creek Valley. It is a little more than a mile long. Its southern end, with the help of equally winding Ridge Road, connects to Beach Drive. At its northern end, it also connects to Beach Drive. In between, there is literally nothing except two rustic picnic areas in the woods. There are no structures or turn offs onto other roads that lead somewhere. It does not provide a shortcut to any other location. The road was constructed just to be a nice, relaxing drive through the woods.

About halfway along this unusually rustic road in the middle of a major city, the road crosses a deep ravine on a high historic bridge. In the words of Spratt, the Ross Drive Viaduct "seems alone far out amid the remote solitude of the forest." Spratt also commented on the tributary crossed by

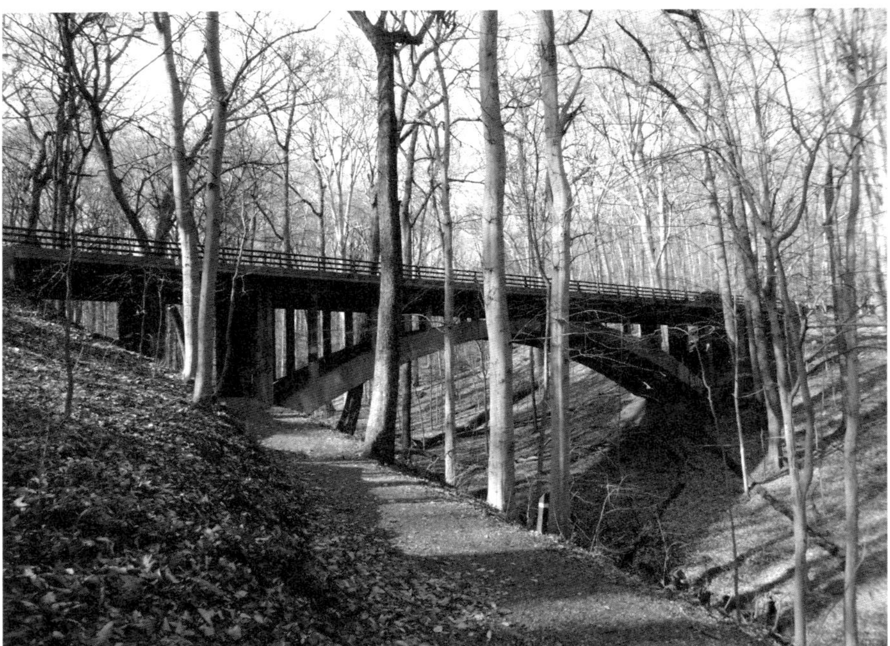

View of the 1907 Ross Drive Bridge in Rock Creek Park.

the bridge. Without offering a name, he referred to it as "a very deep ravine that has a trickle of a brook running down it," and "even when it is at flood stage anyone can easily step across it." Even with no creek, Spratt's deep, steep-sided ravine still required a bridge to maintain a level roadway. A wooden trestle, constructed by chain gangs, was built in 1903. This was later replaced by the current bridge in 1907.

The bridge is an open-spandrel concrete arch with no decoration or ornamentation. It is forty-five feet high and about 170 feet long. Driving across, it is almost unnoticeable. It carries one lane in each direction, with no signs or other indications that it might be an interesting bridge. It has no sidewalks, which is not a problem for visitors because the bridge gets little traffic. There is a small pull-off on its northern end, large enough for two or three cars.

Despite its remote location and few visitors, the Ross Drive Bridge is considered by many authors to be one of the District's most historically important early bridges. It was one of the first triple-hinged concrete arch bridges constructed in the nation, and it is impressively large when viewed in profile. The historic importance of the bridge was recognized when it was placed on the National Register of Historic Places in 1980.

Although the bridge does not cross over a road or substantial creek, it does cross a rustic trail that leads down the slope to Rock Creek. This trail can be used to get some distance from the bridge in order to view and photograph its appealing yet simple structure.

PINEY BRANCH BRIDGE

Similar to Connecticut Avenue, Sixteenth Street NW serves as a major thoroughfare carrying commuting traffic from the Maryland suburbs into downtown. While Connecticut Avenue serves this function for areas west of Rock Creek Park, Sixteenth Street opened up areas for development on the park's east side. Also similar to Connecticut Avenue, Sixteenth Street crossed a deep ravine carrying a tributary, Piney Branch, to Rock Creek. Piney Branch is an east-to-west-oriented tributary entering Rock Creek from the east, almost directly opposite where Klingle Stream enters Rock Creek from the west. The Piney Branch Bridge, also known as the Sixteenth Street Bridge, was completed in 1909 and carries Sixteenth Street across Piney Branch and the Piney Branch Parkway.

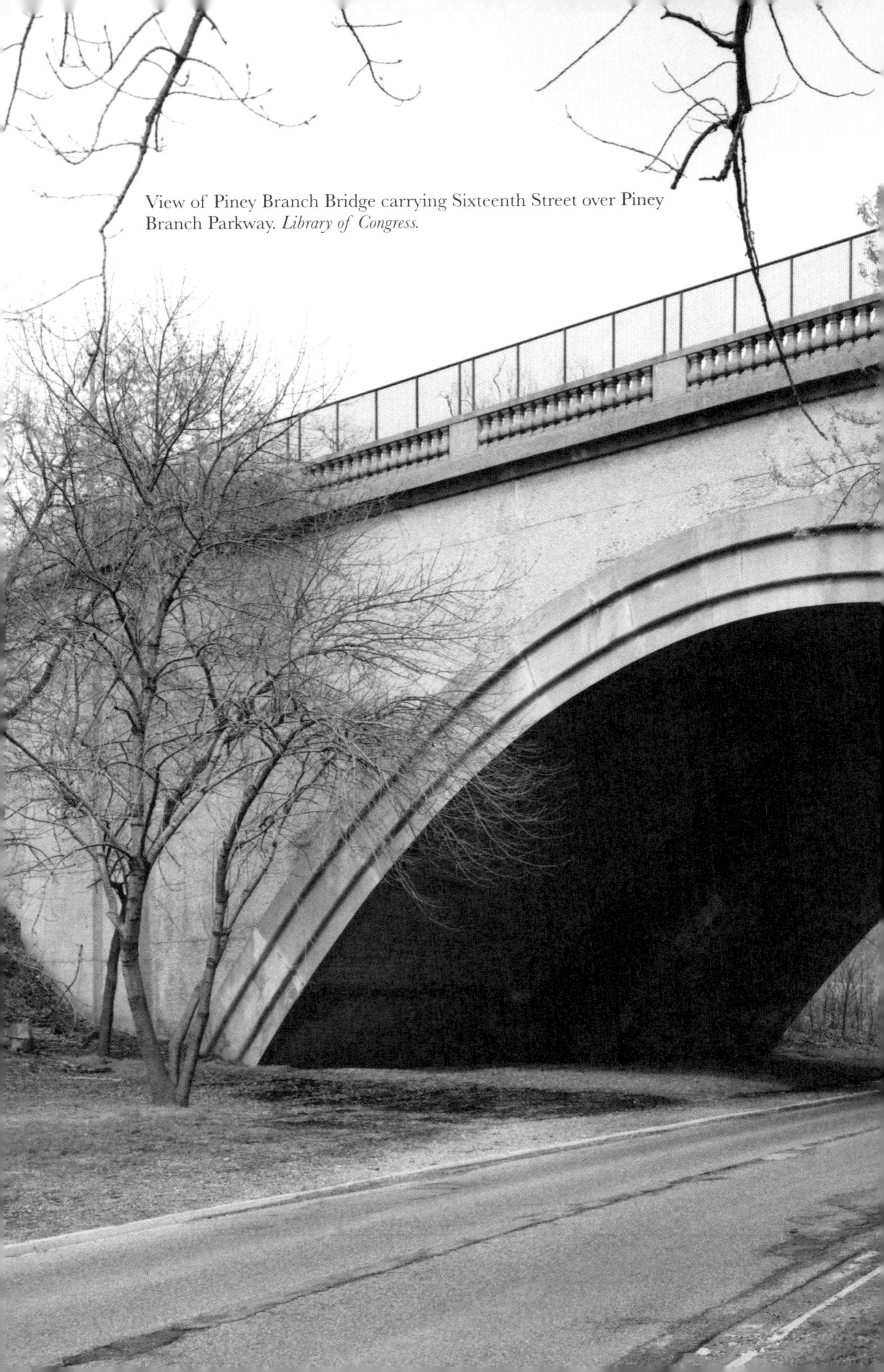

View of Piney Branch Bridge carrying Sixteenth Street over Piney Branch Parkway. *Library of Congress.*

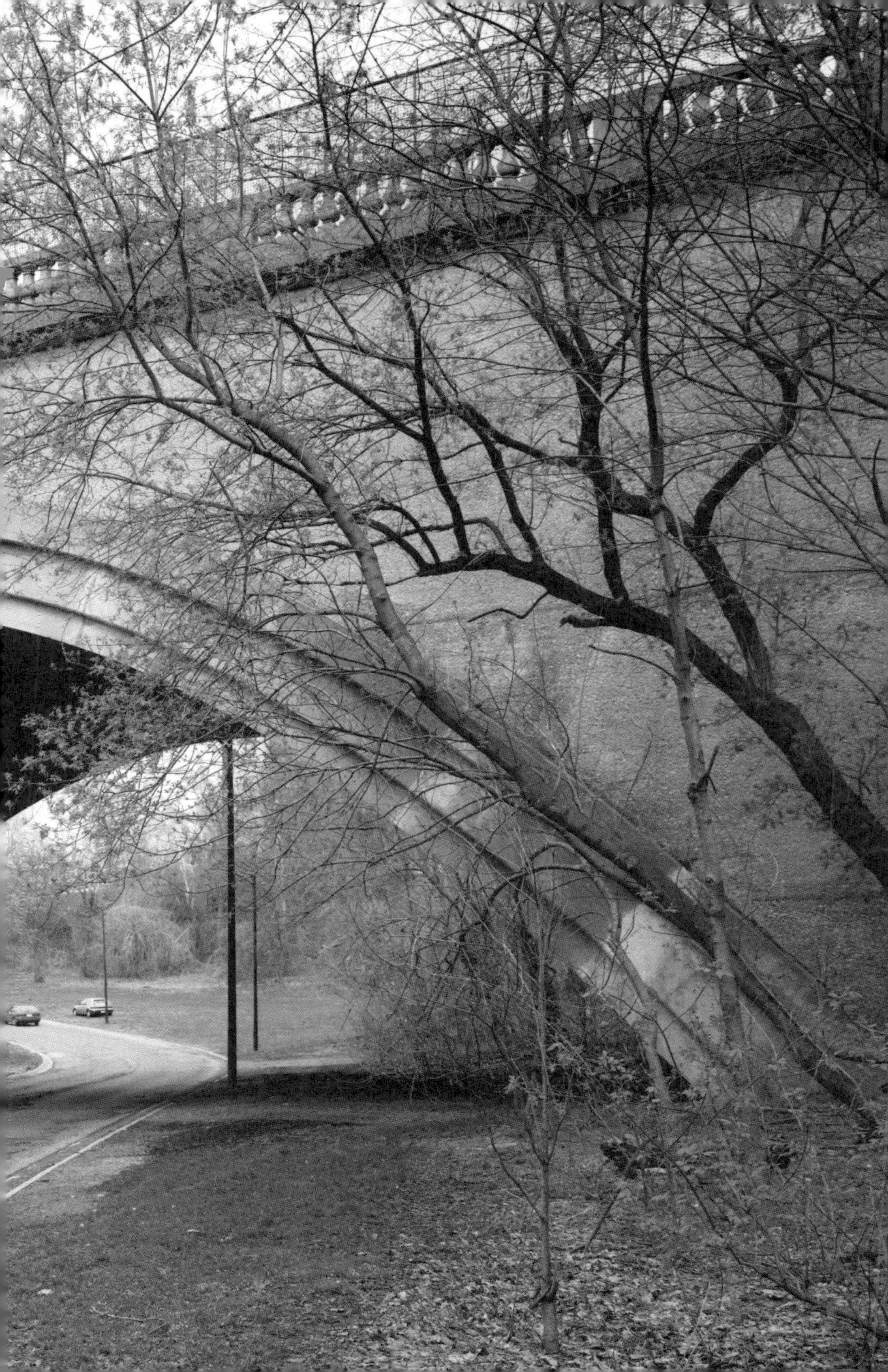

Sixteenth Street NW was proposed as one of the principal roadways of the new Federal City in the L'Enfant Plan of 1791. It extended directly north from L'Enfant's proposed location for the President's House and was the site of some of the earliest commercial and residential development in the city. However, L'Enfant's Plan extended only about two miles north of the President's House and did not include an extension of Sixteenth Street as far north as Piney Branch.

In the nineteenth century, urban development generally followed Fourteenth Street rather than Sixteenth Street, because the Sixteenth Street route was too muddy. As shown by the topographic map of the District from 1900, Sixteenth Street terminated south of the zoo. Fourteenth Street, two blocks east of Sixteenth Street, traversed due north but then veered west and crossed Piney Branch on a bridge in line with Sixteenth Street. This bridge had stone abutments and a wooden superstructure.

It was not until the early 1900s that the current configuration of bridges and roadways in this area was developed. In 1899, Congress authorized the northward extension of Sixteenth Street as a major thoroughfare into Maryland. This required purchasing land, razing structures, straightening Fourteenth Street, and linking the existing bridge to Sixteenth Street instead of Fourteenth.

Developing Sixteenth Street into a major artery also required replacing the old wooden bridge with a modern bridge, which is the current Piney Branch Bridge. The bridge consists of two adjacent concrete arch bridges, fifteen feet apart, the first constructed in 1907 and the second completed in 1909. Also in 1907, Congress authorized the construction of a parkway adjacent to Piney Branch within the ravine as a separate unit of the larger Rock Creek Park. However, the Piney Branch Parkway was not funded and constructed until the 1930s, when New Deal–era funding was provided for infrastructure projects.

The bridge, consisting of the two adjacent, single-span unreinforced concrete arches, was the first parabolic arch bridge constructed in the United States. The spandrel is closed, with a triple arch ring giving three-dimensional depth to the profile. A thirty-six-inch water main is carried between the two arches.

Like the Taft Bridge, different mixtures of concrete and aggregate were used to create visual definition of the various components of the structure. The cast concrete of the balustrade is made of an aggregate of crushed granite pebbles, which was then brushed to resemble masonry before it hardened. The bridge carries five traffic lanes with two sidewalks and measures almost three hundred feet long, with a sixty-foot-high deck above the valley floor below.

You can see the bridge's profile when driving on Piney Branch Parkway beneath the arches or walking along the roadside trail. The parkway is about a mile long and extends from Arkansas Avenue on the east to Beach Drive on the west, intersecting with Beach Drive about halfway between Peirce Mill and the zoo.

Most District residents and commuters rarely see the bridge's profile, but the decorations along the roadway are a landmark for commuters on Sixteenth Street. Like the Taft Bridge, constructed at about the same time, the sidewalks are lined with decorative features. The concrete balustrade is cast into a series of neoclassical balusters, and the ornate lampposts consist of a fluted cast-iron column topped by three white-glass globe light fixtures.

Most striking are the concrete monuments on the four corners of the bridge. Though on a smaller scale than the lions on Taft Bridge, the monuments display gilded bronze statues of tigers, lying with their enormous paws extended in front of them like sphinxes. Alexander Phimister Proctor,

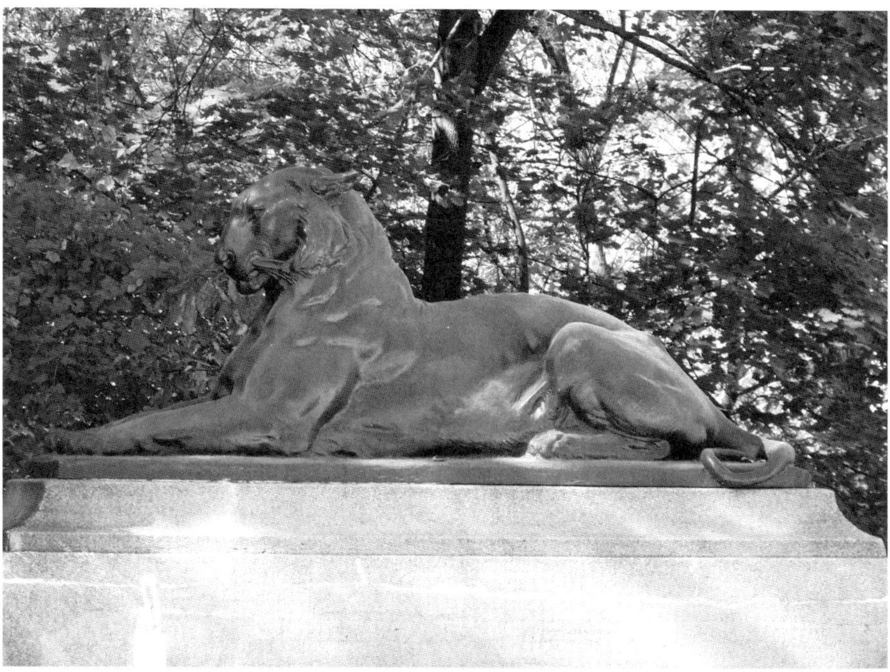

One of four tiger sculptures visible along Sixteenth Street on the corners of the Piney Branch Bridge.

who a few years later would create the bison statues for the Dumbarton Bridge, sculpted the tigers. To complete these statues, Proctor studied tigers at the Bronx Zoo. The statues led to the local name of "Tiger Bridge."

The bridge's sidewalks can be accessed by using on-street parking in the Petworth, Sixteenth Street Heights, and Columbia Heights neighborhoods. The trail under the bridge is most easily accessed from its eastern end at Arkansas Avenue. It can also be accessed from its western end at Beach Drive, but that route requires a hike of almost a mile uphill along the Piney Branch Parkway.

DUMBARTON OAKS PARK AND MONTROSE PARK BRIDGES

One of the more unusual tributaries to Rock Creek is the stream flowing through Dumbarton Oaks Park and Montrose Park. Unnamed in most reference materials and maps, the stream is referred to as "The Branch" in the National Park Service Historic American Buildings Survey report for Dumbarton Oaks Park. Landscaped into a faux-rustic wilderness between 1923 and 1933, the stream within the park is no longer free-flowing and instead consists of a series of small pools separated by decorative waterfalls over stone weirs.

Although all are generally within the Dumbarton neighborhood, Dumbarton Oaks Park, Dumbarton Oaks Gardens, Dumbarton Bridge, and Dumbarton Street are all different geographic features in different locations, which makes it easy to confuse them. The Dumbarton area, sitting on the high bluffs between Rock Creek and Georgetown, was one of the earliest neighborhoods settled in the District. As a suburb separated from the governmental center of the city by Rock Creek, the neighborhood remains one of the more affluent parts of the city.

Two of its early residents established grand estates adjoining each other, and each is surrounded by acres of woods and gardens. The sixteen-acre estate established by Richard Parrott in 1804 was eventually purchased by the U.S. government in 1911 and developed into Montrose Park, another of the dozens of small units encompassed within the larger Rock Creek Park.

The adjacent estate that later became known as Dumbarton Oaks was purchased by Mildred and Robert Woods Bliss in 1920. They hired Beatrix Farrand, a prominent landscape designer, and spent the next ten years developing the gardens and adjoining woods that encompassed The Branch, which enters Rock Creek below the Massachusetts Avenue Bridge.

Farrand and the Blisses made one substantial decision that confuses visitors today. They developed the gardens directly surrounding the mansion into terraces with delicately manicured flower beds, arbors, and gazebos. However, they also transformed the adjoining slopes and creek, consisting of twenty-seven acres, into a more natural, rustic retreat with trails, waterfalls, and bridges. In 1940, the Blisses donated the mansion and the surrounding Dumbarton Oaks Gardens to Harvard University, which still owns and operates them as a museum and research center today. However, they donated the rustic twenty-seven-acre part of the estate to the National Park Service, which administers it as Dumbarton Oaks Park, another of the smaller parks incorporated into the larger Rock Creek Park. The historical importance of the Dumbarton Oaks Park's landscaping was recognized by its placement on the National Register of Historic Places in 1967, with an updated nomination with detailed descriptive information filed in 2004.

The similarity in names is not the only source of confusion. The three separate properties, all open to the public, adjoin each other in a complicated way. Dumbarton Oaks Museum and Garden and Montrose Park are both walled, with designated entrances on R Street in the Dumbarton neighborhood. However, they are separated by a steep, narrow hiking trail that also intersects R Street. This was a former road called Lover's Lane. Although it is within Montrose Park, Lover's Lane also serves as the main entrance into Dumbarton Oaks Park. Despite sharing a name and a long property boundary, there is no direct access from Dumbarton Oaks Museum and Gardens into Dumbarton Oaks Park. They are different properties and are separated by a fence.

This explanation is necessary because Dumbarton Oaks Museum and Gardens does not have any tourist bridges. Similarly, Montrose Park has one footbridge that, while attractive, is not historical. However, Dumbarton Oaks Park has one important and attractive tourist bridge, as well as several more recent replicas and small footbridges that support the use of the hiking trails.

There are about seven "bridges" accessible from the Lover's Lane entrance, depending on how you count them. Six are within Dumbarton Oaks Park, and one is in Montrose Park. The National Park Service inventory of historic bridges in Rock Creek Park lists four in Dumbarton Oaks Park: Clapper Bridge, Clapper Bridge Falls, Gray Memorial Arbor Bridge, and Stone Bridge. However, only two, Stone Bridge and Gray

Memorial Arbor Bridge, are easily correlated to existing bridges in the park. The two references to "clappers," which were early bridges formed by laying flagstones horizontally across stones set on end, are not as obvious. In general, the historic documentation on the park focuses on the different weirs constructed to create waterfalls, but the bridges seem to be an afterthought.

The main attraction is Stone Bridge, the largest and most decorative of the bridges. This is the only bridge shown on the map provided to visitors at the park entrance. Crossing the stream just north of Lover's Lane, the Stone Bridge consists of a stone arch crossed by a flat deck with arch-shaped stone parapets. The bridge is about thirty feet long and fifteen feet wide, and the arched parapets reach a height of about thirty-two inches at the center of the bridge. The stones are generally rectangular in shape yet are irregular enough to appear rustic. Stones placed on edge form a decorative arch ring, and irregularly shaped flat stones provide capping for the arched parapets. The bridge is adjacent to the Old Stone Pump House, a stone structure that is currently used as a pesticide shed. Of all the bridges in the park, the Stone Bridge is where you will see visitors take selfies and wedding photos on sunny spring days.

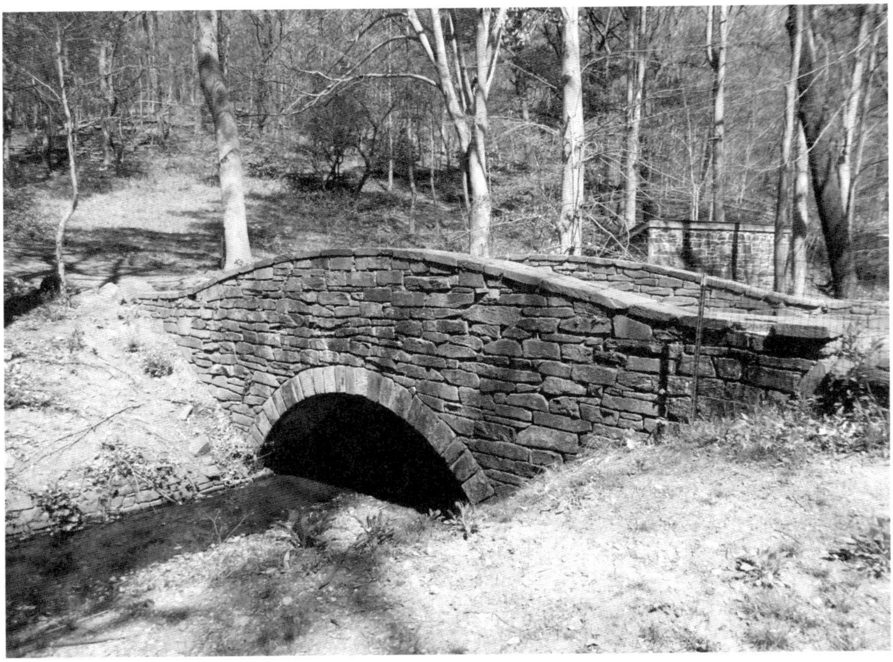

Stone arch bridge in Dumbarton Oaks Park.

Falling into disrepair, but formerly an important focus within the park, is the Gray Memorial Arbor. This structure was constructed sometime between 1937 and 1942 to commemorate William Gray, the estate's first grounds superintendent. The structure originally consisted of a three-sided stone enclosure with wooden benches and a slate memorial plaque, open toward the stream and garden on Forsythia Hill within the Dumbarton Gardens. The plaque no longer exists. The "bridge" here is nothing more than a half-hewn log, flat side up, extending between two stone parapets. The current log was placed in 2000, along with a similar log "bridge" at a location called Clapper Bridge Falls.

The other notable bridge is in the northern woodland section of Montrose Park. Referred to as the "Bridge Over The Branch," this bridge is located where Lover's Lane crosses the stream one final time before reaching Rock Creek. This is a recent bridge, only about fifteen feet long, that was constructed when the trail was rerouted. The bridge consists of steel girders held between stone abutments, but has attractive wooden railings constructed in a geometric X-shaped pattern.

9

THE FUTURE

As discussed throughout the book, the District's bridges are a mixture of historically important locations, a few unattractive disappointments, some pleasantly unexpected surprises, hidden gems, and above all, incredible pedestrian and bicycle access to the riverfronts, neighborhoods, monuments, and tourist sites across the city. But the most important observation regarding the city's bridges is that they are about to get much better.

One of the major themes of *Bridgespotting: A Guide to Bridges that Connect People, Places, and Times* was that, after decades of being dominated by unattractive bridges without any pedestrian access or recreational amenities, bridge construction in the United States turned a corner in the 1990s and early 2000s. For a variety of reasons discussed in that book, it is now almost unthinkable to plan the construction of a new bridge without considering its impact on the visual aesthetics of the city, its commemoration of local cultural leaders, and its contribution to the walkability and recreational infrastructure of the community.

Washington, D.C., was not a national leader in this transition. In fact, after a construction frenzy of characterless bridges as part of the development of the interstate highway system in the 1950s and 1960s, bridge construction in the District went to sleep for almost fifty years. Although the current Ethel Kennedy Bridge, constructed in 2004, was designed to include accessible sidewalks, these were not linked to the regional Anacostia bike trail system until 2016. The Woodrow Wilson Bridge, which incorporates multiple tourist and recreation-friendly features, opened in 2007. However, it sits miles away from the city's center and is not visible from anywhere except the extreme southern tip of Hains Point. The Eleventh Street Bridge was reconstructed in 2013 and included a wide sidewalk with unusual observation platforms sitting on the piers of the previous bridge. However, it is not visually distinctive, especially from a distance.

It was only in 2021 that the lid was blown off with the construction of the new Frederick Douglass Bridge in the South Capitol area, a booming part of the Washington, D.C. real estate market. This was not merely the first bridge in the District with a decorative superstructure designed as a public artwork. Instead, it was the first District bridge that had a superstructure of any kind since the original suspension version of Chain Bridge ceased to exist in 1852. Between the destruction of Chain Bridge in 1852 and the completion of the high white arches of the Douglass Bridge in 2021, every bridge constructed in the District had been made as low-profile and as inconspicuous as possible. While great strides had been made by adding sidewalks to connect trail systems, the Douglass Bridge completed the Anacostia River Trail by creating a comfortable pedestrian avenue between the Anacostia neighborhood and the booming South Capitol area, with its Major League Baseball and soccer stadiums.

The Frederick Douglass Bridge is just the beginning. Plans have been approved, and construction is about to begin on at least three additional major bridge projects that will continue this trend toward community-focused bridges.

NATIONAL ARBORETUM PEDESTRIAN BRIDGE

A new pedestrian bridge across the Anacostia at the National Arboretum is one of the more exciting developments in the city's twenty-first-century bridge construction plan. While the District has done a great job providing functional sidewalks on its large-scale traffic bridges, this will be the first bridge over one of the two major rivers dedicated for pedestrian and bicycle use only.

The location for this new connection fills a lengthy gap between bridges on the Anacostia. Currently, cyclists and pedestrians can cross the river at the Bladensburg Waterfront or at the Ethel Kennedy Bridge at Benning Road, with a bridgeless distance of almost four miles between them. Residents in Deanwood and other neighborhoods east of the river have long wanted a pedestrian and bicycle connection into the city without needing to travel miles south to Benning Road. The bridge will extend a trail to the little-used eastern gate of the National Arboretum at the Asian Collections and then south of the arboretum into the city.

Construction is expected to begin in 2026 and will take up to two years to complete.

ELEVENTH STREET BRIDGE PARK

In chapter 5, the discussion of the Eleventh Street Bridge pointed out that, when the local component of the bridge was replaced in 2013, the piers of the previous bridge were left in place and capped with observation platforms. While the platforms offer attractive views of the Anacostia and are certainly appreciated by the biking and pedestrian communities, they were never intended to be permanent. They were placeholders, while the plans and funding for their next life were developed. If the piers had not had platforms, they would have been eyesores in a red-hot real estate market for at least a decade. The temporary platforms made the appearance of the old piers less obtrusive, while also making them functional.

In the spring of 2026, DDOT plans to begin the contracting process for construction of the Eleventh Street Bridge Park, a partnership between DDOT and a nonprofit called Building Bridges Across the River. To be constructed on the old piers, the park will be much more than a pedestrian and bicycle bridge between Anacostia and the Navy Yard. The bridge will have a unique sculptural X-shaped design that will complement the modernistic design of the nearby Yards Park and Frederick Douglass Bridges. Inspired by the High Line in Manhattan, the surfaces of these X-shaped platforms will provide space for an amphitheater, an environmental education center, a hammock grove, and landscaping with paths among native trees, all in the middle of the river.

LONG BRIDGE RECONSTRUCTION

The 1904 Long Bridge is a major constriction point in the rail system along the entire East Coast of the United States. It was constructed during a time when railroad transport of coal and raw materials into the city was needed to support manufacturing and other industries while, at the same time, daily railroad transport of workers from the suburbs into the city was unknown. Now, 120 years later, the functions of the bridge have been reversed. There is no longer any substantial industry requiring rail transport into the city, while the number of rail commuters from the suburbs has exploded.

The constriction is caused by the fact that there are two separate bridges crossing the Anacostia for freight and passenger trains from the north and east, but the same freight and passenger trains must share a single bridge when crossing the Potomac to the south. On the eastern side of the city, Amtrak and MARC commuter trains cross the Anacostia into Maryland on the Amtrak Rail Bridge adjacent to New York Avenue, while freight trains cross the Anacostia on the Anacostia Rail Bridge south of the Capitol. Within the city, the freight, Amtrak, and Virginia Railway Express (VRE) commuter lines converge to cross the Potomac into Virginia on the Long Bridge. Because all freight trains traveling between the southeastern and northeastern United States must cross both the Potomac and the Anacostia, the need to share the Long Bridge with Amtrak and VRE trains causes severe delays in the regional movement of freight.

Since 2011, studies and plans for refurbishing and updating the Long Bridge have been ongoing. To eliminate the conflicts between passenger and freight trains, a new parallel, two-track bridge for the Amtrak and VRE trains will be constructed, allowing the 1904 Long Bridge to be used only for freight. The plan also includes a parallel pedestrian and bicycle bridge, the first pedestrian-only bridge to cross the Potomac. Ground was broken on the new bridges in October 2024, and completion is expected in about 2030.

SELECTED BIBLIOGRAPHY

Bryan, Wilhelmus Bogart. *A History of the National Capital: From Its Foundation Through the Period of the Adoption of the Organic Act.* Vol. 1, *1790–1814*. Macmillan, 1914. https://archive.org.

Bushong, William. "Glenn Brown and the Planning of the Rock Creek Valley." *Washington History* 14, no. 1 (Spring/Summer 2002): 56–71.

———. *Rock Creek Park Historic Resource Study*. U.S. Department of the Interior National Park Service, 1990. https://archive.org.

Choukas-Bradley, Melanie. *A Year in Rock Creek Park, The Wild, Wooded Heart of Washington, DC.* George F. Thompson Publishing, 2014.

Cohen, Robert. "History of the Long Railroad Bridge Crossing Across the Potomac." National Railway Historical Society, Washington DC Chapter. http://dcnrhs.org.

Cooper, Rachel. "Washington DC Bridges: A Guide to River Crossings." Trip Savvy, June 26, 2019. https://www.tripsavvy.com.

Croggan, James. "'Old Burnt Bridge.'" *Evening Star*, July 7, 1907. www.bytesofhistory.com.

DeFerrari, John. *Capital Streetcars: Early Mass Transit in Washington, D.C.* The History Press, 2015.

———. *Streets of Washington* (blog). www.streetsofwashington.com.

DeFerrari, John, and Douglas Peter Sefton. *Sixteenth Street NW, Washington, DC's Avenue of Ambitions*. Georgetown University Press, 2022.

District Department of Highways. "Washington's Bridges: Historic and Modern, A Pictorial Report on Highway Bridges and Structures in the District of

Columbia." Developed for the Commissioners of the District of Columbia. March 1, 1956.

Dryden, Steve. *Peirce Mill: Two Hundred Years in the Nation's Capital*. Bergamot Books, 2009.

Einberger, Scott. *A History of Rock Creek Park: Wilderness and Washington, D.C.* The History Press, 2014.

Emery, Fred A. 1938. "Washington's Historic Bridges." *Records of the Columbia Historical Society* 39 (1938): 49–70.

Fogle, Jeanne. *A Neighborhood Guide to Washington, D.C.'s Hidden History*. The History Press, 2009.

Gutheim, Frederick, and Antoinette J. Lee. *Worthy of the Nation: Washington, DC, from L'Enfant to the National Capital Planning Commission*. 2nd ed. Johns Hopkins University Press, 2006.

High, Mike. *The C&O Canal Companion: A Journey Through Potomac History*. Johns Hopkins University Press, 2015.

Horne, Robert C. "Bridges Across the Potomac." *Records of the Columbia Historical Society* 39 (1956): 249–58.

Kennicott, Philip. "Washington's Rising Frederick Douglass Bridge Lifts Human Spirit." *Washington Post*, April 13, 2021.

Krakow, Jere L. *Historic Resource Study, Rock Creek and Potomac Parkway, George Washington Memorial Parkway, Suitland Parkway, Baltimore-Washington Parkway*. U.S. Department of the Interior National Park Service, 1990. https://archive.org.

Lewis, Tom. *Washington: A History of Our National City*. Basic Books, 2015.

Moore, Charles, ed. *The Improvement of the Park System in the District of Columbia, Report of the Senate Committee on the District of Columbia (The McMillan Plan)*. Government Printing Office, 1902. https://www.nps.gov.

Muller, Charles G. *The Darkest Day: The Washington-Baltimore Campaign During the War of 1812*. University of Pennsylvania Press, 2003.

Myer, Donald Beekman. *Bridges and the City of Washington*. U.S. Commission of Fine Arts, 1974.

Peck, Garrett. *The Potomac River: A History and Guide*. The History Press, 2012.

Short, C.W., and R. Stanley-Brown. *Public Buildings: A Survey of Architecture of Projects Constructed by Federal and Other Governmental Bodies Between the Years 1933 and 1939, with the Assistance of the Public Works Administration*. U.S. Government Printing Office, 1939.

Smith, Kathryn Schneider, ed. *Washington at Home: An Illustrated History of Neighborhoods in the Nation's Capital*. Johns Hopkins University Press, 2010.

Spratt, Zack. "Rock Creek's Bridges." *Records of the Columbia Historical Society* (1953/56): 101–34.

Sullivan, Francis P. "Concrete Forms and Concrete Surfaces." In *The Architectural Annual*, edited by J.P. Hynes. Architectural League of America, 1907.

Thelwell, Yvonne. "Washington, D.C. District Department of Transportation." *Aspire*, Spring 2022.

U.S. Commission of Fine Arts and Office of Archeology and Historic Preservation, National Park Service. *Georgetown Historic Waterfront, Washington, D.C., A Review of Canal and Riverside Architecture*. 1968.

U.S. Department of the Interior National Park Service. Historic American Buildings Survey (HABS) and Historic American Engineering Record (HAER) Reports for:
Arlington Memorial Bridge (1988)
Bluffs Footbridge (1992)
Boulder Bridge (1988)
Boundary Bridge (1992)
Bridge at High Street
Calvert Street (Duke Ellington) Bridge (1992)
Dumbarton Oaks Park (1994)
Grant Road Bridge (1988)
Joyce Road Bridge (1988)
K Street Bridge (1992)
Klingle Valley Bridge (1992)
L Street Bridge (1992)
Lyon's Mill (Devil's Chair) Footbridge (1992)
M Street Bridge (1992)
Massachusetts Avenue (Charles C. Glover) Bridge (1992)
Milkhouse Ford (1992)
Morrow Drive Bridge (1992)
New Harvard Street Bridge (1995)
Old Harvard Street (1992)
P Street Bridge (1992)
Peirce Mill Bridge (1992)
Pennsylvania Avenue Bridge (1992)
Pinehurst Bridge (1988)
Potomac Aqueduct (1983)
Q Street (Dumbarton) Bridge (1992)
Rapids Footbridge (1988)
Riley Springs Footbridge (1992)
Rock Creek and Potomac Parkway (1992)
Rock Creek and Potomac Parkway Bridge Near P Street (1988)

Rock Creek Park Road System (1998)
Rolling Meadow Footbridge (1992)
Ross Drive Bridge (1988)
Saddle Club (Shoreham Hill) Footbridge (1992)
Shoreham Hill Bridge (1988)
Sixteenth Street Bridge Spanning Piney Branch Parkway (1992)
Tidal Reservoir Inlet Bridge (1988)
Tidal Reservoir Outlet (1988)
William Howard Taft Bridge (1992)
Woodley Lane Bridge Abutment (1992)

U.S. Department of the Interior National Park Service. National Register of Historic Places Nomination Forms for:
Arlington Memorial Bridge (1979)
Boulder and Ross Drive Bridges (1976)
Chesapeake and Ohio Canal (1979)
Dumbarton Oaks Park (2004)
Francis Scott Key Bridge (1995)
Georgetown Historic District (2003)
Godey Lime Kilns (1973)
Jones Point Lighthouse and District of Columbia South Cornerstone (1980)
Main Sewerage Pumping Station (2012)
Q Street (Dumbarton, Buffalo) Bridge (1973)
Rock Creek Park Historic District (1992)
William Howard Taft Bridge (2003)

U.S. Department of the Interior National Park Service. "An Administrative History—Roads and Trails." https://www.nps.gov.

———. *Montrose Park, Cultural Landscape Report*. August 2004. https://www.nps.gov.

———. "National Register of Historic Places Registration Form, Rock Creek Park Historic District." 1991.

Van Dyne, Larry. "Washington, DC: A City of Bridges." *Washingtonian*, March 1, 2006. https://www.washingtonian.com.

Wilson, William H. *The City Beautiful Movement.* Johns Hopkins University Press, 1989.

BRIDGE MAPS

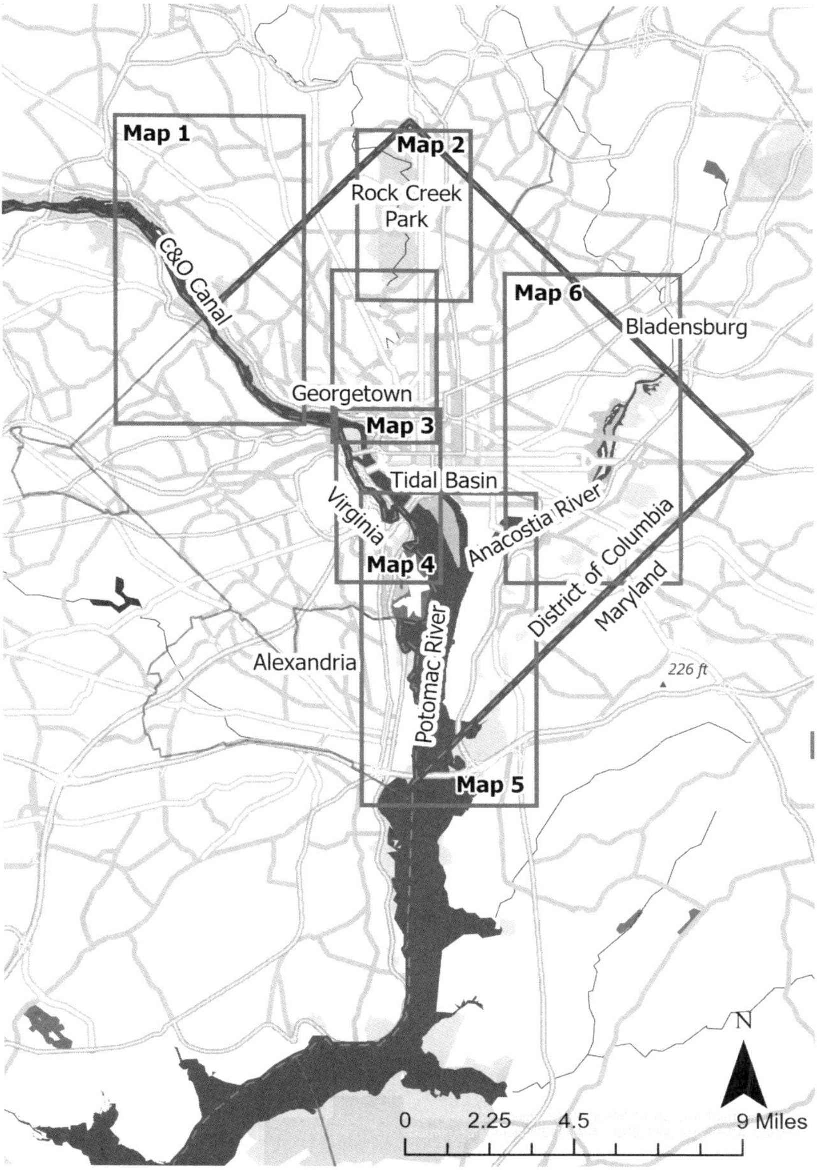

Index to Bridge Location Maps

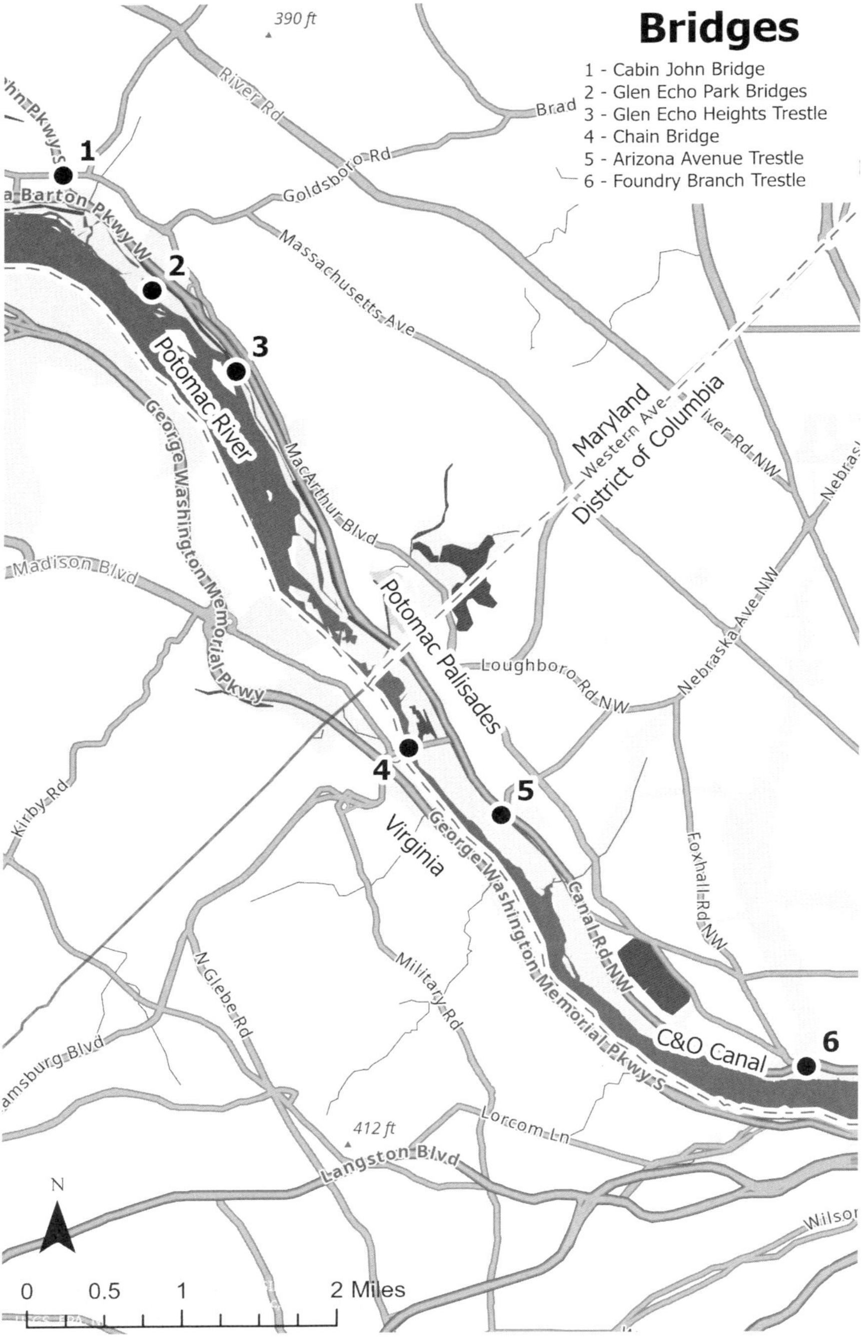

Map 1. Bridge Locations on Potomac River and C&O Canal.

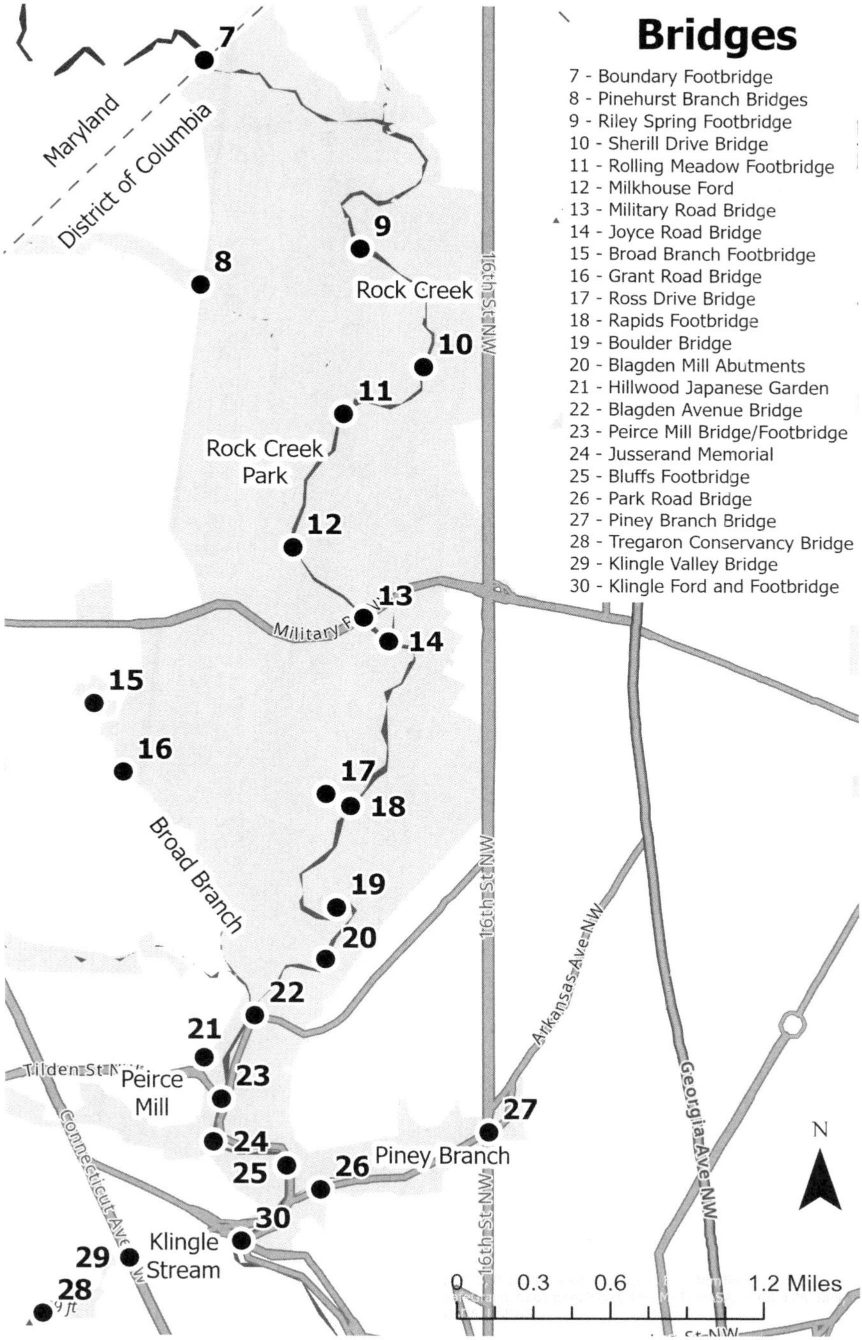

Map 2. Bridge Locations in Rock Creek Park and Rock Creek Tributaries.

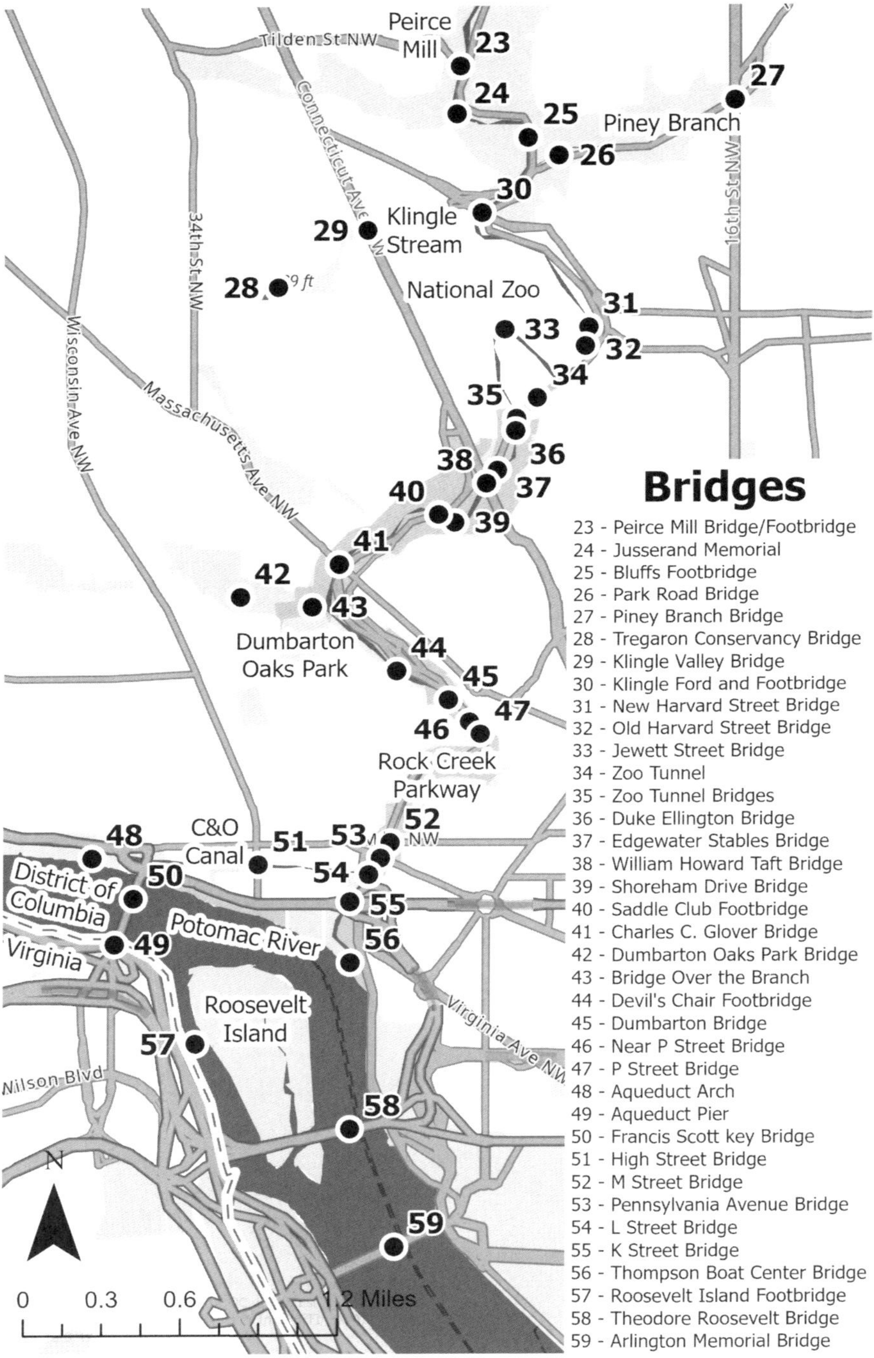

Map 3. Bridge Locations along Rock Creek and Potomac Parkway and on Potomac River and C&O Canal near Georgetown.

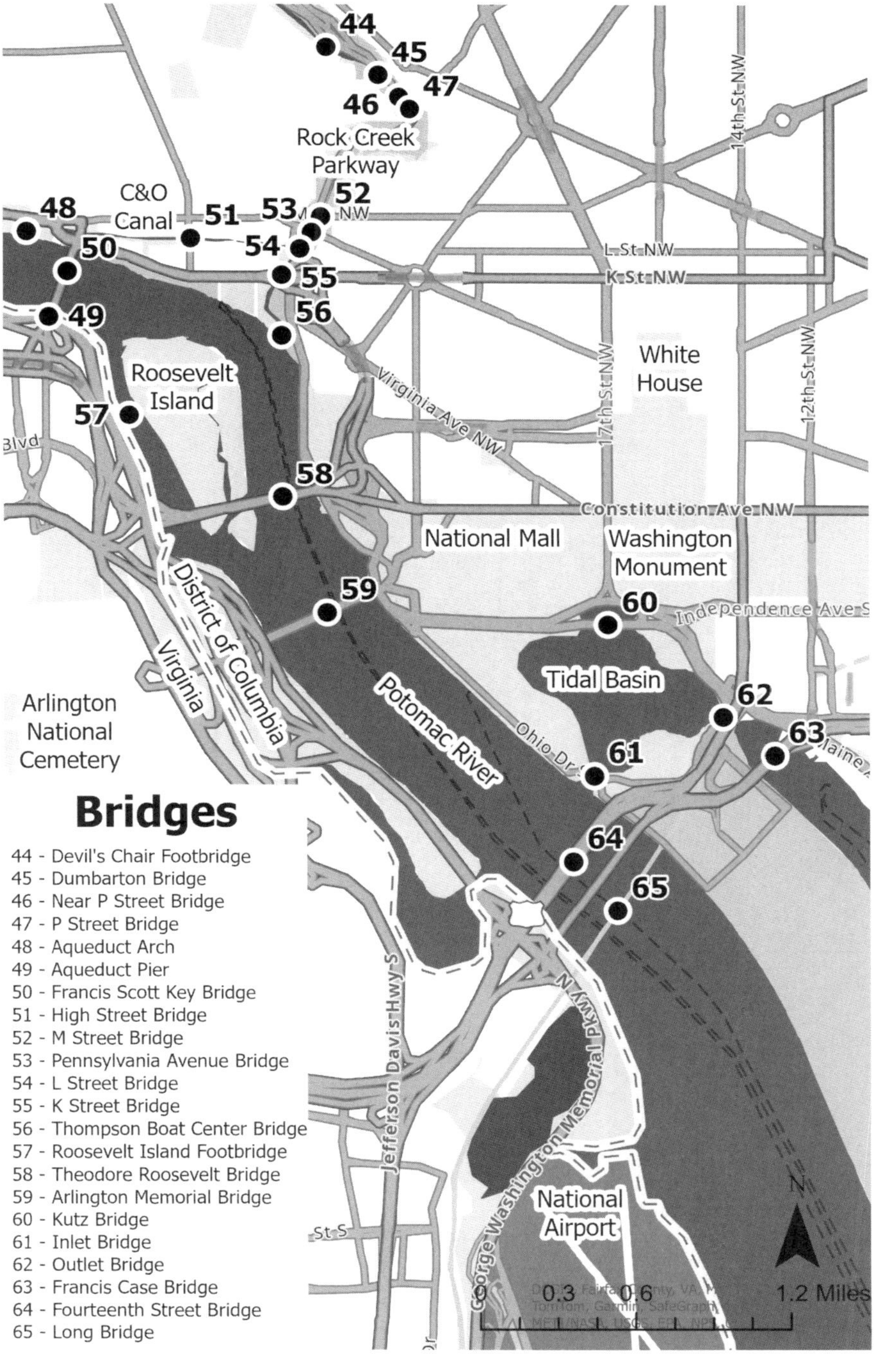

Map 4. Bridge Locations between Georgetown and the Fourteenth Street Bridge complex.

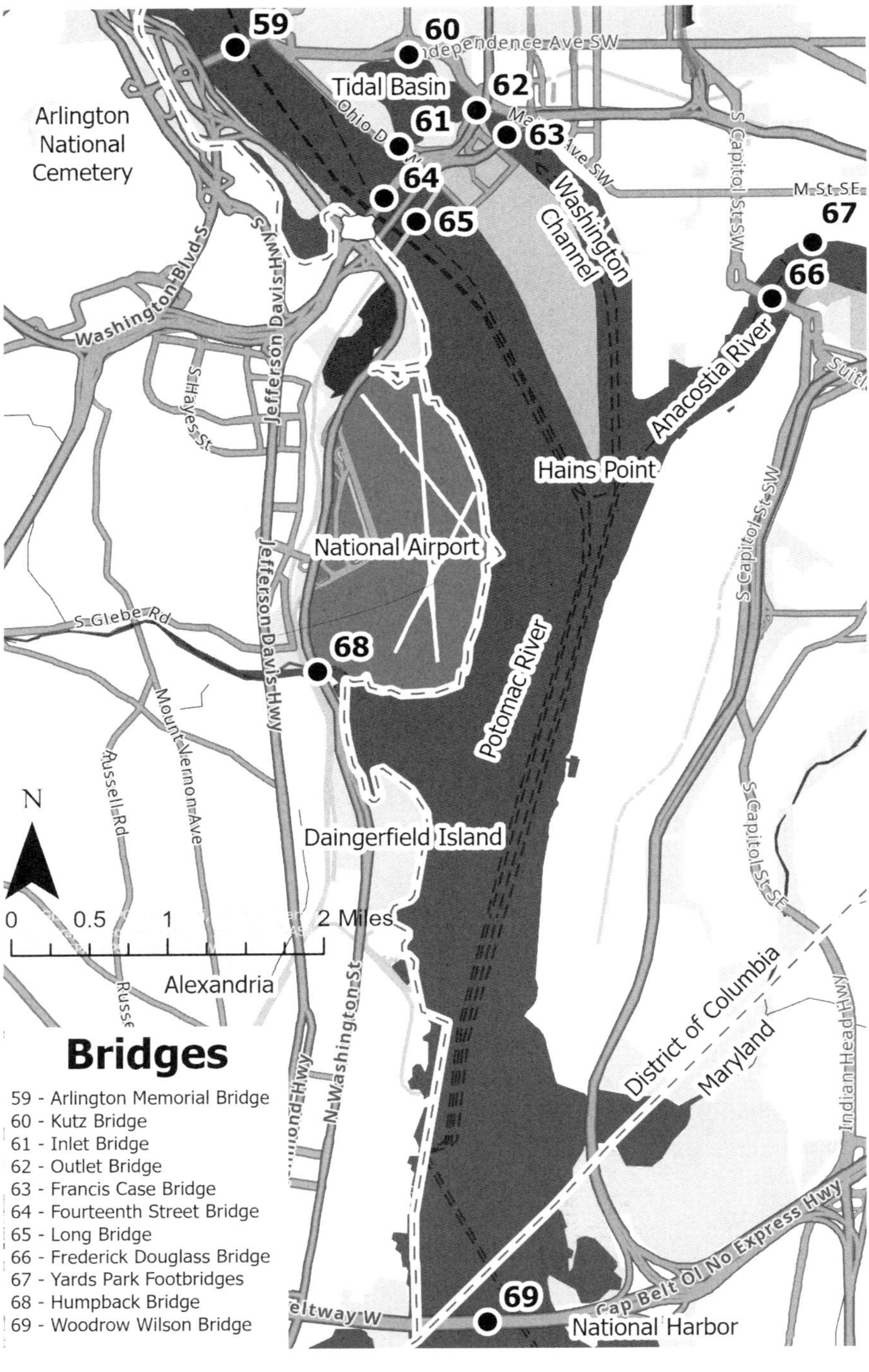

Map 5. Bridge Locations on Lower Potomac and Anacostia Estuaries.

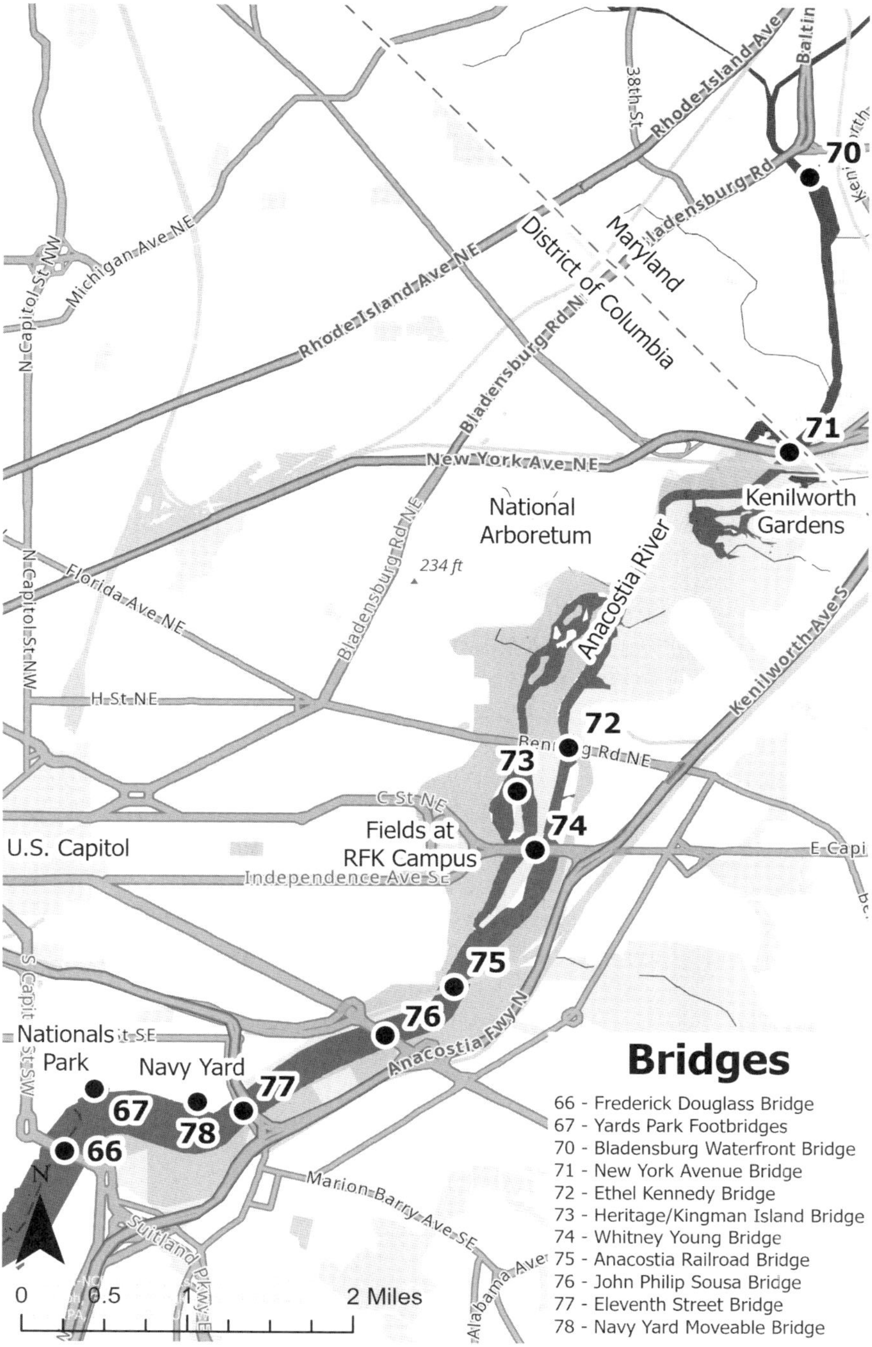

Map 6. Bridge Locations on Anacostia River.

ABOUT THE AUTHOR

Bob Dover is a geologist with thirty-six years of experience in environmental impact assessment. A Washington, D.C. native, he is a graduate of Beloit College and the University of North Carolina. His primary geologic interest is the role of geology and topography in influencing human settlement patterns, architecture, and historical events. In 2022, he published *Bridgespotting: A Guide to Bridges That Connect People, Places, and Times*, a book that examines uses of bridges for tourism and recreation. He is a contributor of bridge photos and documentation to the historicbridges.org website and an active participant in the Vermont Covered Bridge Society and New York State Covered Bridge Society. Recently retired, he lives just outside of Washington in Columbia, Maryland.